Design Considerations for the Space Elevator Apex Anchor and GEO Node

International Space Elevator Consortium
Spring 2017

Authors:
Michael "Fitzer" Fitzgerald
Vern Hall
Peter A Swan
Cathy W Swan

Design Considerations for the Space Elevator Apex Anchor and GEO Node

Copyright © 2017 by:

Michael "Fitzer" Fitzgerald
Vern Hall
Peter A Swan
Cathy W Swan

All rights reserved, including the rights to reproduce
this manuscript or portions thereof in any form.

Published by Lulu.com

pete.swan@isec.org

978-1-387-02452-0

Cover Illustration:
Front – Michael "FItzer" Fitzgerald
Back – ISEC Logo

Printed in the United States of America

Preface

The International Space Elevator Consortium vision is to have:

> "A World with inexpensive, safe, routine, and efficient
> access to space for the benefit of all mankind."

One of the principle elements of our action plan towards an operational space elevator is to conduct year-long studies addressing critical topics. This year, ISEC chose to address the design considerations for the Apex Anchor and Geosynchronous Earth Orbit (GEO) Node. As was discussed in the Architectures and Roadmap report, ISEC understands where the technologies are today and where we would like them to be in order to reach Initial Operational Capability (IOC). The goal of this study team is to add to the "body of knowledge" relative to the two topics addressed herein. This will help us stay focused on our destination as stated below.

Our Destination is IOC for a Space Elevator Transportation System:

The Space Elevator Transportation System is comprised of one Earth Port with two tether termini, an Apex Anchor supporting two 100,000 km Tethers, 14 Tether Climbers, and a single Headquarters and Primary Operations Center. The GEO Node supports the Space Elevator Transportation System with a range of "overhead' functions; e. g. test, safety, and support.

The authors of this report wish to thank:
- Members of ISEC for their support
- Contributors to this report for their dedicated efforts
- Attendees of the 2016 International Space Elevator Conference

Signed: *Peter A. Swan, Ph.D.*
President ISEC

Executive Summary

The study team took on the challenge of expanding the "body of knowledge" pertaining to the Apex and Geosynchronous elements of the Space Elevator. A special effort was made to describe the elements in the context of an entire Space Elevator. The report complements and expands earlier ISEC Space Elevator reports on the Climber, Operations Concept, Architectures & Roadmaps, and Earth Port. To do so, we have established:

1. A robust set of definitions within the Geosynchronous and Apex Regions
2. The Strategic Approach to Architectural Development of a Space Elevator

The Definitions

To ensure complete understanding during this study report, the following definitions are provided:

Space Elevator Column: The volume swept out during normal operations starting at the Earth Port [a circular area within which it operates] and extending through the GEO Region up to the Apex Region. This column of space will be monitored, restricted, and coordinated with all who wish to transverse the volume. The current concept is similar to the FAA's Automatic Dependent Surveillance – Broadcast approach. Satellites, aircraft and ships will announce where they are and coordinate motion through the space elevator column. Each space elevator has a column of allocated volume.

Earth Port Region: The volumetric region around each Earth Port to include a space elevator column for each tether and the space between multiple tethers when they operate together. The Earth Port Region will include the vertical volume through the atmosphere up to where the space elevator tether climbers start operations in the vacuum and down to the ocean floor.

Earth Port: A complex located at the Earth terminus of the tether to support its functions. These mission elements are spread out within the Earth Port Region. When there are two or more termini of tethers, the Earth Port reaches across the region and is considered one Earth Port. [for full definition and explanation of Earth Ports, see both ISEC reports on Architectures and Earth Ports.]

GEO Region: The Space Elevator GEO Region encompasses all volume swept out by the tether around the Geosynchronous altitude, as well as the orbits of the various support and service spacecraft "assigned" to the GEO Region. When two or more space elevators are operating together, the region includes each and the volume between elevators.

GEO Node: The complex of Space Elevator activities positioned in the Space Elevator GEO Region of the Geosynchronous belt; directly above the Earth Port. There will be several sub nodes; one for each tether, one for a central main operating platform, one for each "parking lot", and others. [note: at the GEO altitude, the GEO Node complexes can maintain their locations naturally within the GEO Region.]

Apex Region: The region around the Apex Anchor is defined by the amount of motion expected at the full extension of the tether. The region is the volume swept out by the end of the tether during normal operations. When two or more space elevators are operating together, the region spreads to the volume between.

Apex Anchor: A complex of activity is located at the end of the Space Elevator providing counterweight stability for the space elevator as a large end mass. Attached at the end of the tether will be a complex of Apex Anchor elements such as; reel-in/reel-out capability, thrusters to maintain stability, command and control elements, etc.. [Note: nothing stays at that altitude unless attached to a tether]

As will be discussed in Chapter 4 of this report, the GEO Node is not necessary for IOC per se. Prior to IOC, the GEO Node will provide "overhead" services such as loading and offloading supplies, servicing tugs and other things. However, The GEO Region is expected to become the centerpiece of Space Elevator activities after IOC.

Beyond Initial Operational Capability (IOC), Space Elevator operations at geosynchronous altitude will be critical for business. It will provide services to a myriad of customers. This report examines the transformation of the GEO Region and the maturation of its several GEO Nodes. This transformation and maturation link ISEC's vision of a revolutionary transportation system and the businesses of Space Enterprise of the mid-Twenty First Century.

The above definitions reveal that the Apex Anchor has not been talked about very much relative to the other elements of Space Elevators. In Chapter 3, this report takes a closer look at its main function as counterweight and other functions necessary for operations. The report then looks at the large number of additional services and functions that might evolve at the anchor. A simple way to look at it is to envision a large transportation end-point enabling on-off movement while ensuring stability of the system. As we see the transformations of the GEO Region to the GEO Node, we see the Apex Anchor maturing; becoming a base for supporting construction/refueling/repair; and, a base for exploration initiatives.

The Strategic Approach

The Strategic Approach is ISEC's guiding theme for the technical development of a Space Elevator. ISEC has spent some time discussing how to turn a long-term vision into a long-term plan. The problem is that a plan usually implies either a specific schedule, a specific budget, or both. We have settled on the notion of "an approach" disdaining budget and schedule specifics. How much and when are exigencies; but, the approach must be immutable.

Conclusions

This study continues the push to improve the body of knowledge for Space Elevators. The authors concluded:

- The deployment and continued stability of the tether are the primary function of the Apex Anchor until IOC. This translates to:
    - a reel in/reel out (or climb up/climb down) capability,
    - a capability to fire thrusters (magnitude and direction) as directed by HQ/POC, and
    - support to customers who leverage the strength of the end-point of this space transportation infrastructure.

- The basic mass buildup for the Apex Anchor will initially be from spent climbers and derelict GEO satellites.

- The GEO Node is expected to become the centerpiece of a Space Port that provides "overhead" services such as repair/assembly, refueling climbers, loading and offloading supplies, servicing tugs and many other functions to a myriad of customers, after IOC.

- Today's technology should suffice to understand the needs of Apex Anchors, GEO Nodes, and their customers. However, the future technological capabilities will indeed enhance capabilities in space, especially at these complexes.

- While one is evaluating and developing the Apex Anchor and GEO Node concepts, one must be cognizant of Earth Port design characteristics. Indeed, there are several parallels within this tremendous space elevator transportation infrastructure between the Earth Port and both the GEO Node and Apex Anchor. One of ISEC's favorite images illustrates this on the cover image.

Table of Contents

1 Introduction

This initial chapter lays out the study with broad strokes while placing it in a historical setting. Several volunteers worked on this report and contributed some ideas that were widely discussed previously but not recorded. While others came up with concepts that were unique and new. The brainstorming started early, was orchestrated during the 2016 International Space Elevator Conference mini-workshop on the topic, and continued until the last word was finalized. This pair of topics was open to serious engineering considerations and major definitional activities. Both topics had been discussed before the study was initiated, but there remained many puzzles to address and answers to propose [there are still a few answers eluding heavy analysis]. After this introductory chapter, there is a chapter dealing with developmental concepts: the strategic approach for space elevators and refinement of the two major segments with multiple definitions, as individual arenas and as integrated segments of a space elevator architecture. Each of the main chapters deals with the individual space elevator segments [Apex Anchor and GEO Node] from early developmental phases through Initial Operational Capability to Full Operational Capability. And finely, the conclusions are presented concisely with additional information in the appendices.

1.1 Historical Perspective

This study will look at the Apex Anchor and GEO Node that are common entities to all three modern-day architectures. Over the years the names have changed; and, the missions and concepts have developed. The flow of space elevator architectural concepts is summarized as:

- In 1960, Yuri Artsutanov presented an approach visualizing how it could be achieved – a big leap from Tsiolkovsky's concept from 1895.
- Then, in 1974, Jerome Pearson resolved many issues with engineering calculations of tether strength needed, and approaches for, deployment. This was once again a leap beyond Artsutanov's work and set the stage for the "modern design" for space elevators.
- Edwards established the current baseline for design of space elevator infrastructures at the turn of the century with his book: "Space Elevators" [2003]. He established that the required engineering could be accomplished in a reasonable time with reasonable resources. His baseline is solid; and, it was leveraged for the next two refinements of this transportation infrastructure concept.
- The International Academy of Astronautics study leveraged Dr. Edwards' design and the intervening ten years of excellent development work from around the globe. Forty-one authors

combined to improve the concept and establish new approaches, expanding the Edwards' baseline.

- The latest version of a space elevator architecture is the view by the Obayashi Corporation. Their concept went beyond robotic baselines and included humans as passengers and operators.

For a full assessment of the current architectures go to the ISEC website [www.isec.org] and read "Status of the Space Elevator Concept During the Summer of 2016." [Ref 5]

1.2   ISEC Study Process:

The International Space Elevator Consortium (ISEC) has developed the process of picking a key topic for in-depth analysis and then conducting a year-long study to assess various aspects of the topic. This focus enables the ISEC to prioritize activities and leverage volunteers with expertise in the chosen fields. The single focus on a topic for a particular year enables the community to bring its strengths together and address the topic at the yearly conference, inside the organization's journal, CLIMB, and magazine, Via Ad Astra, and through the study process with a resulting report. The topics chosen by the Board of Directors of ISEC have been:

2010 – Space Elevator Survivability, Space Debris Mitigation
2011 – Carbon Nanotube Developmental Status
2012 – Space Elevator Concept of Operations
2013 – Design Considerations for the Tether Climber
2014 – Space Elevator Architecture and Roadmaps
2015 – Design Considerations for the Earth Port
2016 - Design Considerations for the Space Elevator GEO Node, Apex Anchor
2016 - Design Considerations for the Space Elevator Communications Architecture.
2017 - Design Considerations for Space Elevator Simulation

Each study goes through a similar process, such as:

August 2015   ISEC selects topic at Board of Directors meeting
               Topic announced at the yearly conference
Aug-Dec 2015   Team formed and initial outline of study topics
Jan-Mar 2016 Specific items discussed, analyzed and studied
Mar-Aug 2016   Paper topics submitted to the ISEC International Conference

August 2016	Focus at space elevator conference on topic Mini-workshop brainstorming initiate feedback. [Appendix C]
Sep-Jan 2017	Study topics drafted as chapters in the report
Jan-Feb 2017	ISEC Review of Final Document
Feb-Mar 2017	Final review with top level peer review
April 2017	Publish Study Report

This report will be available on the ISEC website in hardback form for sale and as pdf, for free. www.isec.org

1.3 Chapter Layout

This study report will follow the format of previous ISEC studies and will represent the work accomplished and the conclusions reached. One of the special aspects of ISEC studies is that the work is accomplished by a diverse set of space elevator enthusiasts with special skill sets related to the study topic. Each study is accomplished in about a year with the objective of having the report available for everyone at the conference the following year. The report is laid out as follows:

Chapter 2 Developmental Concepts: This chapter identifies definitions of where we are headed, what the major segments of a space elevator architecture are and lays out a concept towards a strategic approach.

Chapter 3 Apex Anchor: This chapter addresses the upper most part of the space elevator and explains how it is accomplished and what its missions will be. The flow of the chapter shows first how the APEX Anchor develops, and then defines its missions at Initial Operational Capability (IOC). The next section evolves the capability of the Apex Anchor until it reaches Full Operational Capability (FOC).

Chapter 4 GEO Node: This chapter recognizes the Geosynchronous Earth Orbit Region around a space elevator tether as more than an empty space surrounding an altitude. The chapter presents many ideas on how the region will evolve in services and functions beyond IOC. The initial responsibilities of the GEO Region will be minimal – essentially supporting the tether climber deployment of GEO missions. Then it will build up to a space elevator GEO Node full of commercial, government and civil missions enabled by assembly, repair and refueling.

Chapter 5 Conclusions: This chapter lays out the conclusions and recommendations from the study members. The consensus of the study team is laid out so that the recommendations can be initiated and the conclusions clearly understood.

Appendix A is the ISEC Vision and Mission
Appendix B is the Acronym List and Terminology List
Appendix C is the minutes from the brainstorming sessions

2   Developmental Concepts

2.1   Introduction

The Apex Anchor and GEO Node are major segments in this complex transportation infrastructure and should be understood as significant entities and integrated components.

2.2   Strategic Approach

As we have done at our previous annual conferences, we conducted brainstorming sessions for each target entity. The contributions from the participants were amazing. Summaries from the sessions are given in Appendix C.

The major feedback from these sessions was that there needs to be a clear distinction between the transportation revolution and the visionary entrepreneurial outcome of such a revolution. The concepts of the Space Elevator Transportation System and Space Elevator Enterprise must be distinguished; and clearly so! We now realize that a Space Elevator Transportation System is the revolutionary core of our vision and the Space Elevator Enterprise encompasses the transportation core and entrepreneurial manifestation of what could be. A strategic approach has emerged;

The Strategic Approach

The Strategic Approach is ISEC's guiding theme for the technical development of a Space Elevator. ISEC has spent some time discussing how to turn a long-term vision into a long-term plan. The problem is that a plan usually implies either a specific schedule, a specific budget, or both. We have settled on the notion of "an approach" disdaining budget and schedule specifics. How much and when are exigencies; but, the approach must be immutable.

The core of our approach is that ISEC must remain committed to the transportation system we envision; yet realize there is a larger goal. Our "strategy" is to link the Space Elevator Transportation System to the Space Elevator Enterprise within a Unifying Vision: the Galactic Harbour.

The Space Elevator Transportation System will be the core, priority construction activity; and, its success will be the foundation of the Space Elevator Enterprise. They will be built in a manner separate from each other but not in isolation. This "separate but not segregated" paradigm establishes both the prioritization and collaboration between and within our near parallel efforts.

2.3   Parallel Concepts

2.3.1   Earth Port Baseline

To ensure understanding of the layout of this report, the readers should start with the concept shown in ISEC Study Report, "Design Considerations of a Space Elevator Earth Port." Several similar images will establish the parallel relationships between the Earth Port, Apex Anchor and GEO Node. The Earth Port [Ref 6], shown in Figure 1, was defined to be an entity which:

- serves as a mechanical and dynamical termination of the space elevator tether, providing reel-in/reel-out capability and position management in order to deal with tension, wind, current and debris avoidance. It may also serve as a satellite terminus platform
- serves as a port for receiving and sending Ocean-going Vessels (OGVs). The OGVs that come and go from the Earth Port will move tether climbers, payloads, supplies and personnel
- provides landing pads for helicopters from the OGVs
- serves as a facility for attaching and detaching payloads to and from tether climbers; and, attaching and detaching climbers to and from the tether
- provides tether climber power for the 40 km above the Floating Operations Platform (FOP)
- provides food and accommodation for crew members as well as power, desalinization, waste management and other support operations.

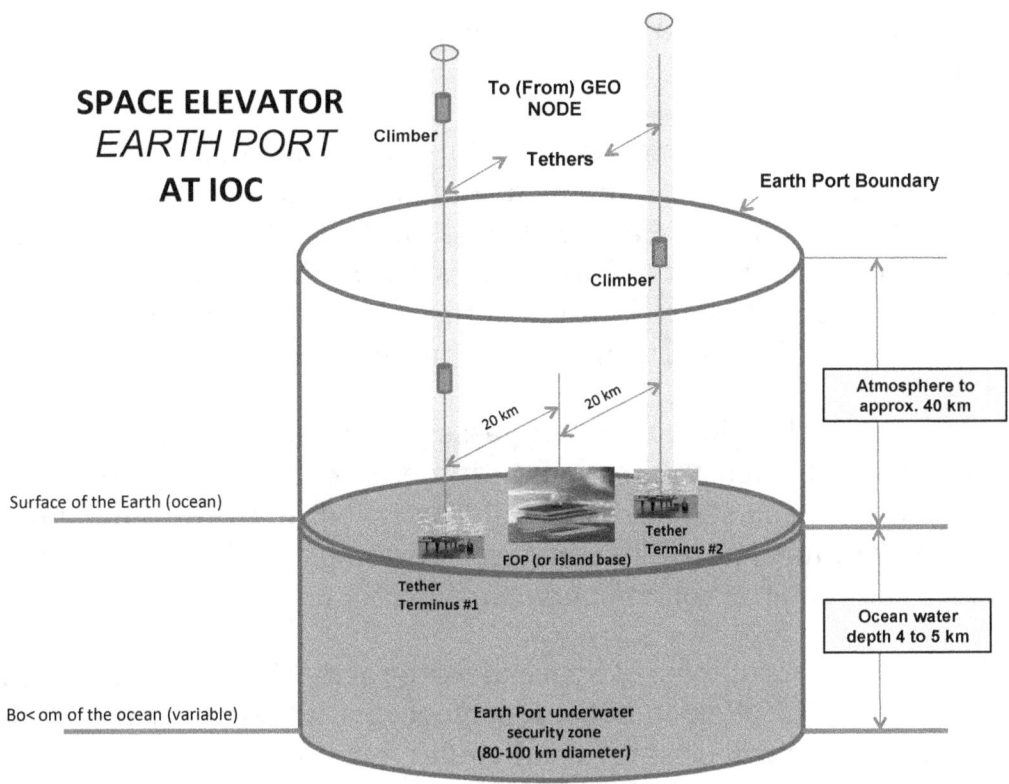

Figure 1 Earth Port Elements

2.3.2 Definitions

To ensure complete understanding during this study report, the following definitions are provided:

Space Elevator Column: The volume swept out during normal operations starting at the Earth Port [a circular area within which it operates] and extending through the GEO Region up to the Apex Region. This column of space will be monitored, restricted, and coordinated with all who wish to transverse the volume. The current concept is similar to the FAA's Automatic Dependent Surveillance – Broadcast approach. Satellites, aircraft and ships will announce where they are and coordinate motion through the space elevator column. Each space elevator has a column of allocated volume.

Earth Port Region: The volumetric region around each Earth Port to include a space elevator column for each tether and the space between multiple tethers when they operate together. The Earth Port Region will include the vertical volume through the atmosphere up to where the space

7

elevator tether climbers start operations in the vacuum and down to the ocean floor.

Earth Port: A complex located at the Earth terminus of the tether and has a complex required to support its functions. These mission elements are spread out within the Earth Port Region. When there are two or more termini of tethers, the Earth Port reaches across the region and is considered one Earth Port. [For full definition and explanation of Earth Ports, see both ISEC reports on Architectures and Earth Ports.]

GEO Region: This Region encompasses all volume swept out by the tether around the Geosynchronous altitude, as well as the orbits of the various support and service spacecraft "assigned" to the GEO Region. When two or more space elevators are operating together, the region includes each and the volume between elevators.

GEO Node: A complex of Space Elevator activities positioned in the Space Elevator GEO Region of the Geosynchronous belt; directly above the Earth Port. There will be several sub nodes; one for each tether, one for a central main operating platform, one for each "parking lot", and others. [note: at the GEO altitude, the GEO Node complexes can maintain their locations naturally within the GEO Region.]

Apex Region: The region around the Apex Anchor is defined by the amount of motion expected at the full extension of the tether. The region is the volume swept out by the end of the tether during normal operations. When two or more space elevators are operating together, the region spreads to the volume between.

Apex Anchor: A complex of activity is located at the end of the Space Elevator providing counterweight stability for the space elevator as a large end mass. Attached at the end of the tether will be a complex of Apex Anchor elements such as; reel-in/reel-out capability, thrusters to maintain stability, command and control elements, etc.. [Note: nothing stays at that altitude unless attached to a tether]

Even though the GEO Node is not necessary for IOC per se, it is expected to become the centerpiece of Space Elevator activities after IOC. Prior to IOC, the GEO Node will provide "overhead" services such as loading and offloading supplies, servicing tugs and many other activities. It will also provide services to a myriad of customers.

Note that there is no defined Geosynchronous part of the Space Elevator addressed in the Architectures and Roadmaps report. No specific functions are needed at geosynchronous altitude; and thus, it is not one of the five segments discussed in that report. Beyond Initial Operational Capability (IOC), Space Elevator operations at geosynchronous altitude will be critical for business.

The above definitions reveal that the Apex Anchors has not been talked about very much relative to the other elements of Space Elevators. This report takes a closer look at its main function as counterweight and other functions necessary for operations. The report then looks at the large number of additional services and functions that might evolve at the anchor. A simple way to look at it is to envision a large transportation end-point enabling on-off movement while ensuring stability of the system.

2.4   Destinations and How to Get There

This section will discuss some of the basic concepts that tie together the space elevator system of systems. First we will show where we are going – destinations. Then we will breakout the various steps on how to do the development of a space system – Sequences. After those two concepts, the flow of the space elevator enterprise will become clearer while helping the understanding of the needs of each of the major segments.

2.4.1 Destinations

The Space Elevator will be a transformational transportation system. It will move objects, systems, material and (eventually) people from the Earth to and from Space. The Space Elevator will be much more efficient than today's launch systems. It will be safe. It will be environmentally friendly, and most importantly, it will enable a wide range of revolutionary activities in space. Because of a Space Elevator, we will do today's space missions better than ever before. It will enable us to do missions in space that we have only dreamt of. One goal is to show the space elevator community's expectations for future missions leading towards an intermediate and then future destinations. As a result, the breakout of a logical flow of capabilities will be shown. As described later in this report, the sequence of development will proceed until the space elevator can robotically move payloads (called Initial Operational Capability – IOC) towards a more robust system performance called Full Operational Capability (FOC).

- The Initial Operational Capability (IOC) consists of a system comprised of two space elevators with one Earth Port with two terrestrial terminus, two Apex Anchors each with 100,000 km tethers, multiple tether climbers and a single Headquarters and Primary Operations Center This system will be capable of moving significant payload tonnage [20 Metric ton] to GEO, and beyond, several times a week from each space elevator.

- The Full Operational Capability (FOC) contains two tethers per elevator system (100,000 km strong tether), each with a tether terminus platform inside the Earth Port, GEO Node, Apex Anchor, and with a single Headquarters and Primary Operations Center. This system will be capable of moving an estimated 70 Metric tons to GEO and beyond several times a week, with passengers.

2.4.2 Developmental Sequences

To set the stage for understanding the development of this unique space transportation system, certain terms have been suggested to express the steps that will be needed to accomplish a full up capability. The following is an assembled set of steps. Each step has an extended phase of tests, simulations and demonstrations. During each developmental phase, there would be many tests, experiments, analyses, and simulations supporting technical maturity leading to deployment and operation of a Space Elevator. The following are the space elevator developmental phases, with explanations following:

1. Pathfinder
2. Seed Tether
3. Single String Testing
4. Operational Testing
5. Limited Operational Capability (LOC)
6. Initial Operational Capability (IOC)
7. Capability On Ramps leading to Full Operational Capability (FOC)
8. Full Operational Capability

1   Pathfinder: In-Orbit Pathfinder Demonstration: This pathfinder is designed as an in-orbit flight demonstration of all possible sub-systems and elements of a space elevator. One essential point is to achieve this early pathfinder in-orbit experiment using near-term technologies – i.e. the tether material need NOT be a full-up CNT ribbon (maybe composed of current Kevlar or beta material of some type).

2   Seed Ribbon Deployment: This component of the Space Elevator will be the basis for a feasible first step in building a space elevator – deployment. The estimate of the technological readiness (in about 2031) will project for a much less capable ribbon being deployed and captured by a "start-up" Earth Port.

3   Single String Testing: In early forms, single string testing could "simply" be an end-to-end simulation of a segment or even the entire architecture. Single string testing is largely investigative, aiding engineering progress and maturation. One single string test could be an examination of the flow of services craft envisioned within the Earth Port or the GEO Node. Single String Testing will continue as the whole tether is reinforced and built up.

4   Operational Testing: OT is that set of test events intended to validate that a system or segment performs as designed in an operational context.   Generally speaking, the tests envisioned here are defined based upon the extension of developmental specifications, the system engineering approach, an overall test plan, and other similar documents.

5   Limited Operational Capability: The idea of LOC is similar to the baseball concept of spring training. All aspects of the Architecture are included and the hardware has been operationally deployed. This phase is good for assessing whether the operating personnel are knowledgeable and trained, that payload customers are aware & understand how this Space Elevator works for them, and operational instruction documents (ie., checklists) are finalized and vetted with "real" operations and operators. This limited capability will be concurrent with the tether buildup activities – adding tether mass from tether buildup climbers.

6   Initial Operational Capability: System engineering competency is part of what IOC is. These "engineering competencies" – validated by execution of the sequenced events – are the functional requirements of

the Space Elevator at IOC. The Space Elevator will function as designed and tested with safety & certainty, be well observed, and in communications contact with HQ/POC. The ISEC also sees the Space Elevator as a valued part of the space business enterprise in the latter part of this century – a useful and valued partner with a wide set of business entities.

7   On-Ramp to FOC: The need for Space Elevator capability growth after IOC should be obvious; but to be clear, the Space Elevator post-IOC on ramp activity will be a formal process by which we add more of the IOC's functionalities, improved versions of the IOC functionalities, and new Space Elevator functionalities. The result will be movement towards Full Operational Capability (FOC) with incremental additions of capability. In practice, on ramp activity is ➔ More; ➔ Better; and then ➔ New.

8   Full Operational Capability: The visionary aspect of the Architecture includes tourism, interplanetary travel staging, hospitals, factories, power generation and a multitude of operational support services. The FOC vision of the Space Elevator will expand with time and be achieved by expansion via the "more", "better", and "new" paradigm cited in the on ramp sequence. The basis of each expansion will be the engineering maturation achieved by progressing through the sequenced steps cited in this paper.

2.5   Prelude

The next two chapters leap from the concepts presented in this preliminary layout to projections of the future. The Apex Anchor is required from the initial deployment of the first tethers and develops into a complex set of regions where the future is very bright. The GEO Region is established with the initial deployment satellites, expanded by corporations and governments dropping off mission satellites and then grows into complex GEO Nodes with robust activities.

3 Apex Anchor

3.1 Introduction

This chapter addresses a Apex Anchor Node of a Space Elevator Transportation System. The 2012 ISEC Study Report, "Space Elevator Concept of Operations," describes missions of the Apex Anchor as follows:

> The Apex Anchor mission is multi-dimensional; but, its principal function is to provide stability for the space elevator as a large end mass. This will ensure a firm tether for the climber, and provide a constant outward force. In addition, the Apex Anchor will have the mission of reeling the tether in and out as required for various tasks such as debris avoidance, damping tether end librations, and reacting to emergencies.

The two definitions applicable to this chapter are:

> Apex Region: The region around the Apex Anchor is defined by the amount of motion expected at the full extension of the tether. The region is the volume swept out by the end of the tether during normal operations. When two or more space elevators are operating together, the region spreads to the volume between.

> Apex Anchor: The complex located at the end of the Space Elevator providing counterweight stability for the space elevator as a large end mass. Attached at the end of the tether will be a complex of Apex Anchor elements such as; reel in/out capability, thrusters to maintain stability, command and control elements, etc.. [note: nothing stays at that altitude unless attached to a tether]

This chapter describes an Apex Anchor and how to develop it into an end mass at an estimated altitude of 100,000 km. A quick look at the Apex Region will set the stage for further discussions. This chapter will then describe apex anchors, discuss general approaches for building or creating them, and illustrate a four-phased approach for development. During the chapter the baseline concept is explained that the Apex Anchor is the original deployment satellite moved to the top at 100,000 km as it extends all of its tether. The mass is then increased to ensure balance with the rest of the tether mass build-up below it. After that, this chapter will initiate the discussion of how to build-up the mass of the tether and Apex Anchor as they are related. The last portion will show the Apex Anchor functions necessary to reach IOC. The last part will focus on customer-driven needs beyond Initial Operational Capability toward Full Operational Capability.

The Apex Region, as been defined above, will resulting from an understanding of the volume swept out by the upper tether terminus. There are a few items that are assumed when discussing a future Apex Region:

- The full volume of this Apex Region is in "hard" vacuum.
- The volume rarely has transits of space rocks or human spacecraft.
- Freedom to operate within the Apex Region will become extremely valuable.
- Estimates of this swept out volume are currently being looked into with the 2017 ISEC Study entitled "Design Considerations for Space Elevator Modeling and Simulation."
- Significant fuel would be required for thrust to keep objects within the region without attachment to the Apex Anchor.

The essential questions with an Apex Anchor is the combination of how big should it become and how will it grow? The Apex Anchor must be there for stability of the entire tether system. Initially, it will provide the separation forces during deployment; over time, it will grow to provide the principal source of stability for a space elevator. Essentially, the Apex Anchor will provide the outward force with tensile pull along the tether leading to inherent space elevator stability. The ratio of mass of an Apex Anchor to the mass of the tether is a mathematical equation that must be calculated and considered during deployment and build-up of the tether. The IOC relationship of masses is approximately 1,900 metric tons for the Apex Anchor and 6,300 metric tons for the tether.

A quick vision of the Apex Anchor was given in the IAA study report [Ref. 1] to help explain the concept of the Apex Anchor as a key segment of a space elevator infrastructure. This vision establishes the total concept well past IOC toward FOC.

> "An Apex Anchor is the upper terminus (counterweight) of the Space Elevator. Apex Anchors will have solar arrays for power and serve as an end platform for climbers and a station for climbers that would descend. Several other activities will occur at the Apex Anchor to include its use as a platform to construct, deploy, recover, maintain, and repair satellites. It will also be a lab to conduct experiments utilizing the low-"g" environment and to analyze the material samples mined and retrieved from the Moon, asteroids, or other planets. At a

later stage, it might be a facility for tourists. The Apex Anchor will also be equipped with facilities to manage tether dynamics, telecommunication, attitude control, collision avoidance of meteorites or space debris, manned activity including EVA, and transfer vehicles. The Apex Anchor will enable construction of space systems and refueling of spacecraft. Operations of the Apex Anchor are controlled at the HQ/POC. Operations will have multiple activities during construction with continuous activity once commercial operations commence. There would be several space elevator center of mass management activities; including, reeling in or out of the tether from the Apex Anchor, thruster control of motion at this upper terminus, and coordination with HQ/POC ensuring a stable upper tether. In addition, the motion for avoidance of space debris could also be controlled or initiated from the Apex Anchor." [Ref. 1]

There are three current concepts for addressing an Apex Anchor that fulfill the developmental needs; however, the third one is preferred by most current developmental architectures.

Approach 1 – No Apex Anchor, Bare Tether
A tether about 150,000 km long will be intrinsically balanced around the GEO altitude, with the tether above balancing the tension from the tether below. No Apex Anchor would be required.

Approach 2 – Edwards' Asteroid
The Edwards proposal, presented in his book [Ref 2], describes the use of an asteroid to serve as the Apex Anchor, as one of the options for end mass.

Approach 3 – Deployment Satellite High
In this proposal, the tether-deployment satellite itself provides the basis for the Apex Anchor from deployment through operation. The satellite begins at GEO and moves outward to counterbalance the descending tether. All three modern day space elevator developments baseline this concept [Ref. 1, 2, & 3]. An example is the Obayashi Corporation approach (Ref 3). They have opted for the deployment satellite to release the tether downward only, with thrusters at both the deployment satellite and the lower end of the tether. This raises the deployment satellite to become the upper terminus creating the initial Apex Anchor baseline mass. Their tether reinforcement satellites run up to GEO and then continue on to become part of their Counterweight [they did not use the term Apex Anchor]. They calculate that there would be 510 individual tether

reinforcement satellites to build up the tether to full capability; then, they would add themselves to the counterweight for mass buildup [Ref. 3].

3.2   Current Apex Anchor Development Concept

3.2.1 Apex Anchor Development Stages

Deployment begins with the deployment satellite being raised to GEO. The Apex Anchor has been discussed at a very cursory level during studies of three major modern space elevator architectures. The discussion below will try to explain the four stages to stabilizing a space elevator with an outward force at its Apex Anchor.

Figure 2   Deployment Satellite Concept
             [image by P. Ellis]

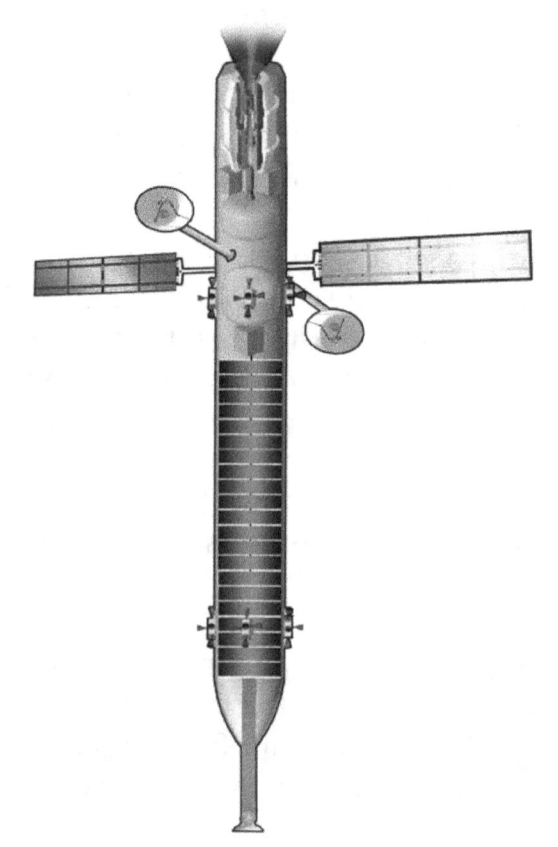

Stage 1 – Initial Deployment: Currently, the only way to get to a GEO orbit for the release of the tether is through rockets with limitations on mass and difficulties in cost and schedule. The first step is to assemble the Tether Deployment Satellite in Low Earth Orbit. This would include multiple launches and assembly on orbit – both capabilities will be improving as the launch opportunity approaches [+/- 15 years]

After assembly in orbit, the large mass [>80 MT] would be raised to a GEO altitude orbital slot with efficient engines that may take up to six months to reach GEO. A concept of this deployment satellite is shown in Figure 1. There is an inherent need for an Apex Anchor for stability. Once the massive satellite reaches its GEO altitude, it will release the single string tether – downward only. This image shows 21 reels, in sequence, providing the 100,000 km of initial (thin) tether. There is a requirement to increase angular momentum with rocket thrust during this deployment activity. This could require refueling at GEO to continue the smooth deployment of the tether. Once the preliminary Apex Anchor has been established with a single string tether to the Earth Port, both the Apex Anchor mass and the strength of the tether must be reinforced.

Stage 2 – Apex Anchor Buildup to Match Tether Mass Buildup: Once the initial portion of the tether has been deployed downward towards its Earth Port terminus, tether stability control is shared between the Apex Anchor and the Earth Port. During the buildup of tether strength and mass toward the Initial Operational Capability, the mass needed at the Apex Anchor must increase consistent with the tether buildup. The approach to add tether mass and Apex Anchor mass in parallel is being discussed with unique ideas and activities required to accomplish construction of a space elevator [see section below for specifics]. Once the stability of the tether is achieved and a sufficient mass/strength of the tether and the Apex Anchor are reached, the development of a second space elevator must be initiated to ensure our continued ability to "beat gravity." This would probably require a second assembly at LEO, movement to GEO and deployment. At the end of the deployment phase, the deployment spacecraft becomes a principle part of the Apex Anchor. This would include computational capability, thruster ability, fuel storage [with refueling capability], and communications links to the HQ/POC, Earth Port, tether climbers, and customers' satellites.

Stage 3 – Initial Operations Capability: The Apex Anchor at IOC will have as its primary mission the stability of the entire tether. This will require the correct mass, thrusters to move the Apex Anchor as needed, reel-in and reel-out capability, as well as the off-load and on-load capabilities retained from the buildup phase.

Stage 4 – Customer Support toward FOC: This phase will be consistent with the growth of the space elevator. Customers will determine where they want their assets delivered and what (if anything) they want to accomplish at the Apex Anchor. This could stretch across the full spectrum of space operations of the future, to include at least refueling, servicing, repair, construction, human habitats and inbound capture as well as routine release into the solar system.

3.3 Space Elevator Mass Buildup

This section expands on the concept of building up the strength and mass of both space elevator tether and Apex Anchor. As they are related, they must be accomplished in parallel. The deployment satellite structure will have an initial mass of approximately 14 metric tons, and an expected required mass of the Apex Anchor at IOC of roughly 1,900 metric tons. This mass will be brought up the tether to the Apex Anchor and then offloaded.. There will be tether climbers designed for this task that will begin during the single string tether developmental phase and continue until at least IOC.

3.3.1 Tether Mass Buildup Methods

The source of mass for the initial space elevator Apex Anchor is the deployment satellite. During the developmental phases of the space elevator, the deployment and buildup of the tether is dominant. The mass at the Apex Anchor for stability must be in a suitable ratio to the mass of the tether, varying with its expected length. The initial seed tether, or single string tether, has a deployment satellite that is very small because the mass must be delivered to GEO by the conventional, costly, method of rockets. The initial single string deployment ends up with a tether mass and an Apex Anchor that are NOT sufficient for IOC or FOC operations. The tether must be built up from this very strong, but embryonic status. The IAA study design showed a mass of 80,000 kg for the deployment satellite with a breakout of approximately half for tether, 17% for satellite and 33% for fuel. As such, the current plan is to strengthen the tether with many "build-up" climbers. These would be small climbers with large masses of tether material. The exact method of deploying and combining new tether to existing tether has not yet been developed; however, the fact is the tether must grow from the minimum needed to be created as a single string to an operational ribbon with characteristics for climbing and survival. The Edwards' book showed that approximately 230 tether climbers must ascend and fuse/meld/ weave/ etc. the enhancement tether sections into the total ribbon[Ref. 2]. The Obayashi Corporation concept use 510 tether buildup climbers[Ref 3]. The final estimate of the IOC tether was given in the IAA study [Ref 1].

> "The conclusion is that the nominal space elevator being discussed
> in this cosmic study has: Tether Mass: 6,300,000kg – with Apex
> Anchor Mass: 1,900,000 kg. " [Ref 1]

3.3.2 Apex Anchor Build-up

In parallel with the growth of the initial tether mass from 40 to 6,300 metric tons, the Apex Anchor must grow from 14 to 1,900 metric tons. The following table shows estimates of the bare tether, single string deployment, and then the IOC option.

18

Approximately	Bare Tether	Initial Single String	Initial Operations Capability (IOC)
Tether Length (km)	152,000	100,000	100,000
Tether Mass (MT)	75	40	6,300
Apex Anchor Mass (MT)	Zero	14	1,900

Table 1 Comparison of Tether Masses

To enable this build-up to IOC with 230 trips of build-up climbers, each tether climber would need to average approximately 25 MT of tether and 8 MT of climber. As such, this study team believes the initial estimates accomplished in 2002 are optimistic by a factor of two – leading to a new estimate of 450 trips for tether build-up climbers. Each of these would consist of 12 MT of tether and 4 MT of climber (on average).

Special Apex Anchor Mass Sources: The following discussion expands upon some important ideas refined during this study. The concept of an Apex Anchor was not developed very extensively during initial discussions of a modern day space elevator. Two of the three architectures [Edwards & Obayashi] used words to describe their Apex Anchor, but did not go into detail on the source of the mass for this counterweight. However, the International Academy of Astronautics spent more time defining what one would look like and how it grows its mass at the 100,000 km altitude. From those discussions, the follow conclusions were detailed:

> "Mass for the counterweight is available from many sources to include near-by asteroids, tether climbers, dead GEO satellites, and additional mass from Earth as required. The initial masses will probably come from the vehicles used to place the initial tether fibers into orbit and from the small tether climbers used in the tether's construction. One source of "extra" mass beyond LEO that can be accessed relatively easily is the set of "dead" satellites at GEO" [Ref 1].

The following discussions are new and follow-up on the IAA study which relates to the latest information gathered by the research study team.

Apex Anchor Growth: After completing their only task, reinforcement of the tether, the climbers are then deemed functionally superfluous and can be moved to the Apex Anchor for extra mass. In addition, if that is an insufficient mass for the ratio to be correct, another approach to adding mass is from retired GEO satellites. The concept would be to deliver defunct satellites to tether build-up climbers that have accomplished its original tasking and then grab, attach it, and take it to the Apex Anchor. There are more than enough derelict spacecraft around the GEO orbit to help fill out Apex Anchors. The concept would be to supplement the mass beyond the climber buildup mass. To expand upon this, the following thoughts are offered:

- There are quite a number of derelict satellites in, or slightly above, the GEO orbit [roughly 400 derelict satellites and 427 active, as of 2016].

- They are owned by someone – in space, it is perceived that there are no salvage rights for those who can get there and move them. Current legal statements explain that even if the spacecraft is inoperable (no energy, no fuel), the responsibility for that satellite still resides with the launching country. [The legal question is up for discussion and could be quite advantageous to the Space Elevator community if a right of salvage is implemented]

- Most derelict GEO satellites are going to eventually approach the space elevator, as they drift after they have run out of fuel. As per direction from the space debris community, for their graveyard phase, each satellite at its end of life should be designed to be placed a reasonable [200 km] distance above the GEO Arc for safe disposal. As such, they will drift backwards toward any space elevator.

- It would be possible to grapple them as they go past and then move them to a location designed to receive and mount them onto a tether climber for travel to the Apex Anchor. [note: it is all downhill from there to the Apex Anchor and would require very little operating power, only very big brakes] If they average around 2000 kg, 1000 satellites should be sufficient to supplement tether climbers for the mass needed (i.e., mass of tether climbers and GEO communications satellites). This activity must be in parallel with the build-up of the tether mass so that the ratio stays relatively constant.

Perhaps, the concept would be to design the tether build-up satellites [mentioned earlier] to have dual purpose; first to bring tether material to the appropriate height and bind it to the existing tether; and then, to go on to the GEO altitude and accept a derelict satellite for delivery to the Apex Anchor. In addition, this activity will clean up the GEO Arc enabling safer operations throughout the region [note: if most of their solar cells were detachable, continued use could be leveraged at the GEO Region].

3.4 Functional Breakout

The introduction of functions for the Apex Anchor in the following sections are shown in two ways, a summary table that shows timelines up to IOC as well as a listing of individual concepts inside an architectural approach defined in the ISEC Study Report, "Space Elevator Architecture and Roadmaps" [Ref 4]. Later, a similar table will be shown for post IOC functions.

3.4.1 Early Development and IOC

The following table has been structured to understand the early missions and functions necessary at the upper end-point of space elevator.

Stages	Support	Description
1, 2	Deployment	Provide support to infrastructure during initial tether deployment.
1, 2, 3, 4	Stability	Continuous control and stability of the Apex Anchor must be maintained. This would start with sufficient mass for appropriate length of the space elevator. It would also include thrust direction and magnitude, reel in and out, and timely knowledge of Apex Anchor locations.
1, 2	Buildup	The obvious activity is the buildup needs strengthening of the tether with additional mass supporting the appropriate tether ratio. However, as the tether gains mass, the need for additional mass at the Apex Anchor becomes critical to keeping the appropriate ratio.
1, 2, 3, 4	Sever Planning	Reaction to sever of tether through multiple pre-planned activities.
Note:		Stages 1 – deployment, 2 – buildup, 3 – IOC, 4 – towards FOC

Table 2 Apex Anchor Functions up to IOC

The ISEC conducted a year-long study, "Space Elevator Architecture and Roadmaps [Ref 4]," looking at key development approaches for the system of systems called the Apex Anchor. The results in that chapter made interesting reading with the study conclusions being shown below. The first four paragraphs deal with the development and activities up to IOC, while the last one shows the move toward FOC.

"Now that the Apex Anchor Segment has been grown from the Deployment Satellite, the actual challenges and testing requirements can be discussed. The major challenges are: Stabilize the initial delivery of the space elevator, establish desired dynamics of the tether, and support customer activities. These are shown in the Roadmap figure shown next."

This process is shown below in the Figure 3 for the roadmap of Apex Anchor development: [Ref 4]

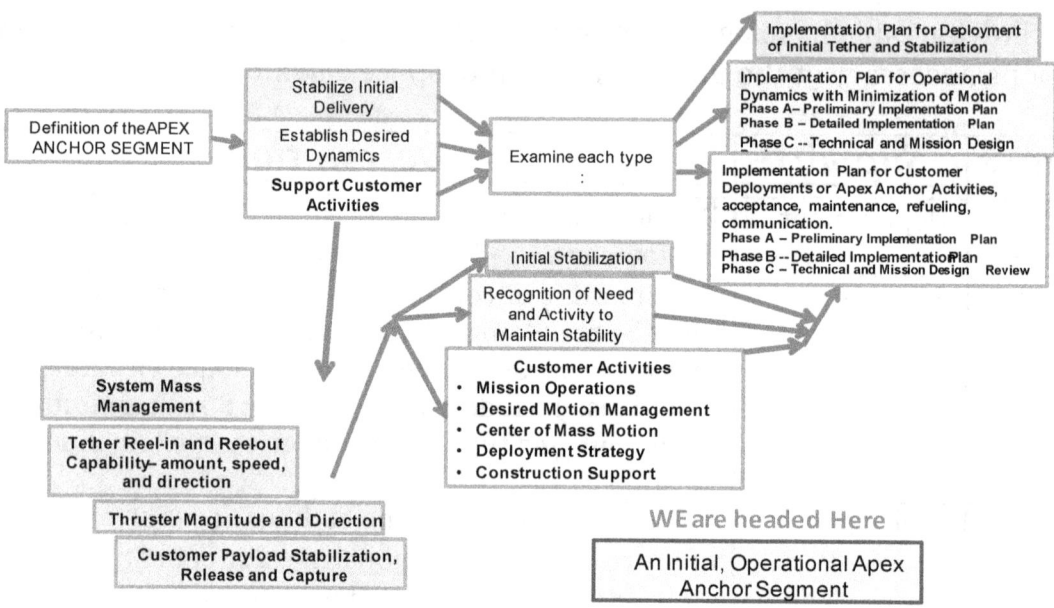

Figure 3 Roadmap for Apex Anchor Segment [Ref 4]

<u>System Mass Management</u>:   The Apex Anchor's principle task is to ensure that the space elevator is stable and providing continuous knowledge of each element of the infrastructure. Critical functions include management of the center of mass of the system as well as understanding the change in mass center or motion as tether climbers are added and moved.

<u>Tether Reel-in and Reel-out Capability</u>:   This task is relatively simple for the Apex Anchor, as it has just released 100,000 km of tether during deployment. The upper end of the space elevator must be able to assist the Earth Port in understanding, recognizing tether element locations, and

controlling every element of the infrastructure. The two segments will work together by reeling-in and reeling-out the tether as needed to ensure the stability of the system.

<u>Thruster Magnitude and Direction</u>: During operations, the motion of the Apex Anchor must be controlled. This will require thrusters and the ability to point them in almost any direction with variable thrust.

<u>Customer Payload Stabilization, Release and Capture</u>: Once the space elevator initiates customer support, the desires will vary. The three approaches that seem to be in demand by potential customers: stable location for mission [stay attached to tether], release to mission trajectory, and capture for placement on a return tether climber.

3.4.2 Expanded Functions at IOC

Indeed, the Architectures and Roadmaps [Ref 4] Study Report showed the missions and functions for the Apex Anchor two years ago. Some of the inherent aspects of the Apex Anchor mission will be to communicate with Headquarters/Primary Operations Center (HQ/POC) to ensure complete knowledge of location and motion of the Apex Anchor. In addition, local activities must be visible to the HQ/POC as well, driving the demand for an excellent communications link. If we expand on each of the functions active up to and including the IOC time period, they would be described as:

- Deployment Support: Provide support to full infrastructure during initial deployment of tether. As the total space elevator is derived from the process of deploying tether and Apex Anchor, this segment's function will dominate during deployment and early buildup phases.

- Stability Support: Continuous control and stability of the Apex Anchor must be maintained. Concepts for control of the dynamics of the tether are expanding as more people look at the problem. Some elements of tether dynamics control are:

- Mass at the upper end provides inherent stability,
- Reel-in and reel-out will provide forces to the tether that can be used to dampen motion
- Thrusting in horizontal directions can dampen motion in the total tether.
- Tether climbers residing at the Apex Anchor, or along the tether, could be leveraged to climb down the tether – thereby putting forces on the tether in a controlled manner to damp out motion.

- Buildup Support: The obvious activity is the buildup and strengthening of the tether with additional mass supporting appropriate tether strength requirements. Support from the Apex Anchor will be in the communications, control and acquisition of any of the tether buildup climbers reaching the Apex Anchor. In addition, there will be the task of off-loading and ensuring safety when mass is delivered to the Apex Anchor, whether it be buildup climbers or derelict satellites.
- Sever Support: Reaction to the severing of a tether must initiate multiple pre-planned activities. The concept being developed relies upon the belief that the space elevator can survive a sever if it is cut in the lower reaches of space. If the tether is cut at the highest danger zone, 800 km altitude, the remainder of the space elevator will react to the loss of mass and connection force. The belief is that with quick reaction at the GEO Node and the Apex Anchor, the total space elevator above that sever altitude might be saved. This would require many actions in a timely manner such as release of tether from both GEO Node and Apex Anchor as well as motion of tether climbers on the remaining tether. Knowledge of the cut must be almost instantaneous [can be accomplished with today's sensors and communications capabilities], while the support must be pre-planned and almost instantaneous.

3.4.3 Evolving towards FOC

This section will address possible Apex Anchor mission functions and what they might contribute to customer needs beyond IOC.

- The Apex Anchor will be a key element in the stability of a space elevator infrastructure during operations.
- The Apex Anchor will ensure support to operations, movement, storage, off-loading, on-loading, placement of customer payloads, etc. In addition, it will help to ensure that space elevator components and scientific equipment are safe, protected, enabled, and robust.
- The Apex Anchor will support activities along the tether itself, especially above GEO, facilitating access to other locations and orbits in CIS-lunar space and beyond. The space elevator provides a "free" toss toward other bodies in our solar system. This release of customer payloads towards destinations around the Moon and within our solar system will be supported by the Apex Anchor with communications links, location determination, last minute operations checks and any real-time emergencies that occur. One aspect of it is that release at roughly 46,745 km in altitude on the space elevator enables capture by the Moon or

placement at the Earth-Moon L-1 location with appropriate velocity corrections at the upper end of a hyperbolic orbit. As such, the robust development of the Earth-Moon ecosphere can be enhanced by a revolutionary delivery technique – sling-shot from a space elevator. By climbing higher, direct tosses toward some of our solar system's planets, especially Mars, becomes routine and inexpensive. The concept is that the Apex Anchor will be the operations focal point for interplanetary operations as well as CIS-Lunar payload movement. This takes advantage of the fact that above the minimal altitude required for a release, the velocity increases linearly. As such, one can arrive faster at the destination – there is no worry about additional fuel to gain extra velocity. All that is required is climbing to a higher altitude on the tether and releasing when the flight path is aligned.

Apex Anchor Development Stage: Coming back to the vision of the CIS-lunar sphere, the necessary support beyond IOC must be expanded. The fourth stage for the Apex Anchor growth is "Customer Support at / near the Apex Anchor towards FOC." This step will have its own characteristics and strengths enhanced by the understanding that it is a key element in space elevator daily operations.

Stage 4 – Customer Support towards FOC: This phase will be consistent with the growth of the entire space elevator infrastructure. The customer will be king and will determine where they want their assets delivered and what they want to accomplish along the tether and at the Apex Anchor. This could stretch across the full spectrum of space operations for the future, to include at least refueling, servicing, repair, construction, human habitats and inbound capture as well as routine releases within and through our solar system.

Functions after IOC: Possible functions for the Apex Anchor are shown in a summary table that crosses the IOC to FOC timelines. Individual concepts are also shown. This chart is one expanded from Table 2; however, the previous emphasis was on the time period up to IOC. This view expands the analyses towards active customer needs that will push development.

Stages	Support	Description
1, 2	Deployment	Provide support to infrastructure during initial tether deployment.
1, 2, 3, 4	Stability	Continuous control and stability of the Apex Anchor must be maintained. This would start with sufficient mass for appropriate length of tehter. It would also include thrust direction and magnitude, reel in and out, and timely knowledge of Apex Anchor locations.
1, 2	Buildup	The obvious activity is the buildup, strengthening, of the tether with additional mass supporting the appropriate tether ratio. However, as the tether gains mass, the need for additional mass at the Apex Anchor becomes critical to maintaining the appropriate ratio.
1, 2, 3, 4	Sever Planning	Reaction to sever of tether through multiple pre-planed activities.
2, 3, 4	Infrastructure	Provide stability, accelerations, power, communications, environmental conditions, security and energy. Tether repair capabilities will enable continuous monitoring and immediate response to issues.
4	Customer	Provide stability, acceleration, power, communications, environmental conditions, security and energy. In addition, provide storage [both open and enclosed] capability for customers.
4	Docking	Provide capability to off-load and on-load systems from tether climbers and/or retrieved satellites. This would include rendezvous with spacecraft coming from CIS-Lunar space as well as up the space elevator.
4	Refueling	Enable refueling of space systems as well as the Apex Anchor itself. This capability could be a commercial mission as well as an aspect of the Apex Anchor capabilities.
4	Human Habitat	Support to humans will be necessary at the Apex Anchor. This will ensure that people will be safe, comfortable, and entertained.
Note:		Stages 1 – deployment, 2 – buildup, 3 – IOC, 4 – towards FOC

Table 3 Apex Anchor Functions

During the following look at missions and functions of the Apex Anchor, the basic concept remains the same. The evolution from pre-IOC will be driven by customer requirements that we can not specify today. However, the following items have been estimated as part of the demands put upon the space elevator community during the transition towards FOC. As seen in the previous section, the developmental approach plans for the continuous improvement of the Apex Anchor from initial stability needs to the support of customer activities. The explanation of the actual functions after IOC start with the basic ones necessary to get to IOC and expand to customer support functions. The bottom line is that the Apex Anchor would continue to provide stability for the space elevator, and support customer needs as they develop beyond IOC towards FOC. Some new functions towards FOC are:

Infrastructure Support: Provide stability, acceleration, power, communications, environmental conditions, security and energy. Tether repair capabilities will enable continuous monitoring and immediate response to concerns. The growing belief within the Space Elevator community is that anything above GEO will be coordinated at least and probably controlled from the Apex Anchor.

Customer Support: Provide stability, accelerations, power, communications, environmental conditions, security and energy. It would seem natural, that when customers arrive at the Apex Anchor, they will expect normal support for their equipment, payloads, satellites, and support personnel. This would definitely include hotel-like accommodations as well as spaceport-like support for operations.

Docking Support: Provide capability to off-load and on-load systems from tether climbers and/or external satellites. This would include rendezvous with spacecraft coming from CIS-Lunar space as well as along the space elevator. As customers approach the Apex Anchor, their expectations will include a haven for their equipment and themselves as well as the simple activity of safely docking and undocking.

Refueling Support: Enable refueling of space systems as well as the Apex Anchor itself. This capability could be a commercial mission as well as an aspect of Apex Anchor capabilities. One of the questions on the table now is where does the fuel come from? It would seem that bringing the fuel up from Earth would be expensive [cheaper than launching of course]. Perhaps the Apex Anchor is a location for acquiring fuel from Lunar or asteroid sources and selling it for a profit at the upper terminus of the space elevator.

Human Habitat: Support to humans will be necessary at the Apex Anchor. This will ensure that humans will be safe, comfortable, and entertained. The concept of a spaceport at the Apex Anchor has many advantages. Not only would it be a safe haven for humans [easy up and down], but it would be the focus for travelers going both ways. A human habitat would enable many functions at that altitude to include construction crews for interplanetary craft, operations personnel for commercial activities at the spaceport, and holiday activities with spectacular views. The advantage of having a small acceleration force inherent in the location could be a tremendous lever for construction or repair of spacecraft occurring in a hanger like spaceport.

The following image shows the Apex Anchor Region in a form that is similar to the Earth Port Region illustration, Figure 1.

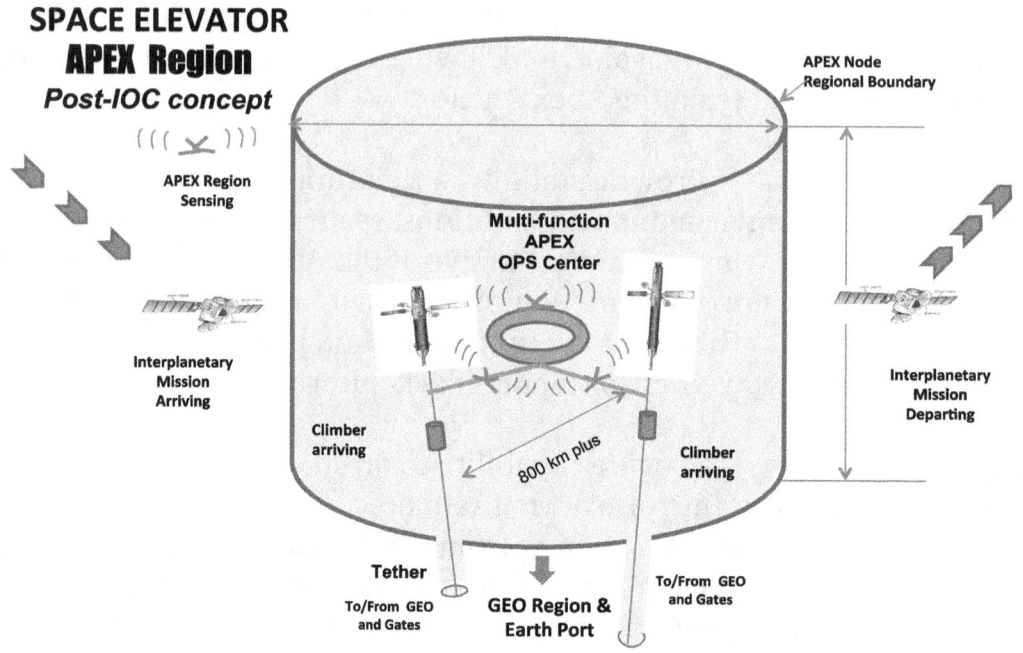

Figure 4, Apex Anchor, Post IOC

3.5   Conclusions and Recommendations

Conclusions:

The deployment and continued stability of the tether are the primary function of an Apex Anchor until IOC. This translates to: a) a reel in/reel out (or climb up/climb down) capability, b) the capability to fire thrusters (magnitude and direction) as directed by HQ/POC, and c) support to customers who leverage the strength of the end point of this space transportation infrastructure.

The basic mass buildup for the Apex Anchor will initially be from spent climbers and defunct GEO satellites.

While evaluating and developing the Apex Anchor and GEO Node concepts, one must be cognizant of the Earth Port design characteristics. Indeed, there are several parallels within this tremendous space elevator transportation infrastructure between the Earth Port and both the GEO Node and Apex Anchor.

Recommendations:

Stimulate further research in development of the Apex Anchor, especially with respect to the accumulation of the masses for both tether growth and Apex Anchor stability.

References:

[Ref 1]   Swan, Peter, David Raitt, Cathy Swan, Robert Penny, John Knapman. *Space Elevators: An Assessment of Technological Feasibility and the Way Forward*, IAA, Virginia Edition Publishing, 2013.

[Ref 2]   Edwards, Bradley and Eric Westling, *Space Elevator – A Revolutionary Earth-to-Space Transportation System*, BC Edwards publishing, 2002.

[Ref 3]   Ishikawa, Yoji, The Space Elevator Construction Concept, Obayashi Corporation, 2013, IAC-13-D4.3.6.

[Ref 4] Fitzgerald, M, R. Penny, P. Swan, C. Swan, Space Elevator Architectures and Roadmaps, ISEC Study Report, lulu.com, 2015

[Ref 5] Swan, P., David Raitt, Space Elevator – 15 Year Update, Journal of British Interplanetary Society, Vol 69, No 06/07, Dec 2016., also on www.isec.org

[Ref 6] Penny, Hall, Fitzgerald, Design Considerations for Space Elevator Earth Port, ISEC Study Report, lulu.com, 2016.

[Ref 7] Penny, Swan, and Swan, Space Elevator Concept of Operations, ISEC Study Report, lulu.com, 2012.

4   GEO Node

4.1   Introduction

As noted earlier in this report, the Space Elevator is sometimes discussed as an innovative transportation system revolutionizing access to space. In other contexts, the Space Elevator is the nucleation point and enabler for a broad and evocative set of enterprises that are the foundation for the betterment of mankind. One simple way to look at it is to consider the missions already at geosynchronous orbit (such as weather satellites and DIRECTV] and multiply them by ten, and then add servicing, refueling, and robotic repair so the missions sustain and improve. Geosynchronous orbit seems the place to be for growth after IOC. ISEC defines a GEO Region; couplet of new terms supplementing each other.

To ensure complete understanding during this chapter, the following definitions are offered:

GEO Region: The Space Elevator GEO Region encompasses all volume swept out by Space Elevator Tethers around the Geosynchronous altitude, as well as the orbits of the various support and service spacecraft "assigned" to the GEO Region. When two or more space elevators are operating together, the GEO Region spreads to include each and the volume between.

GEO Node: The complex of Space Elevator activities positioned in the Space Elevator GEO Region of the Geosynchronous belt; directly above the Earth Port. There will be several sub nodes; one for each tether, one for a central main operating platform, one for each "parking lot," and others.

Space Elevator Column: The volume swept out during normal operations starting at the Earth Port [a circle of area within which it operates] and extending through the GEO Region up to the Apex Region. This column of space will be monitored and restricted with coordination for all who wish to transverse the volume. The column concept is similar to the FAA Automatic Dependent Surveillance – Broadcast approach. Satellites, aircraft and ships will announce where they are and coordinate motion through the space elevator column. Each space elevator has a column of allocated volume.

Early on in our considerations of the design of a GEO Node, ISEC knew that the GEO Node offered no active primary function to the Space Elevator's role re: transportation access to space. Simply stated, the Space Elevator's climber could pass through GEO on its trek to the APEX. In that sense, GEO is only a destination. The GEO location is valuable "only" because that is where initial customers will release their satellites. Thus, the GEO Node functional makeup was originally conceived as minimal; two tether columns passing through and a GEO Node Operating Platform (GNOP) for overhead functions (shown in a slightly inclined orbit). See Figure 5.

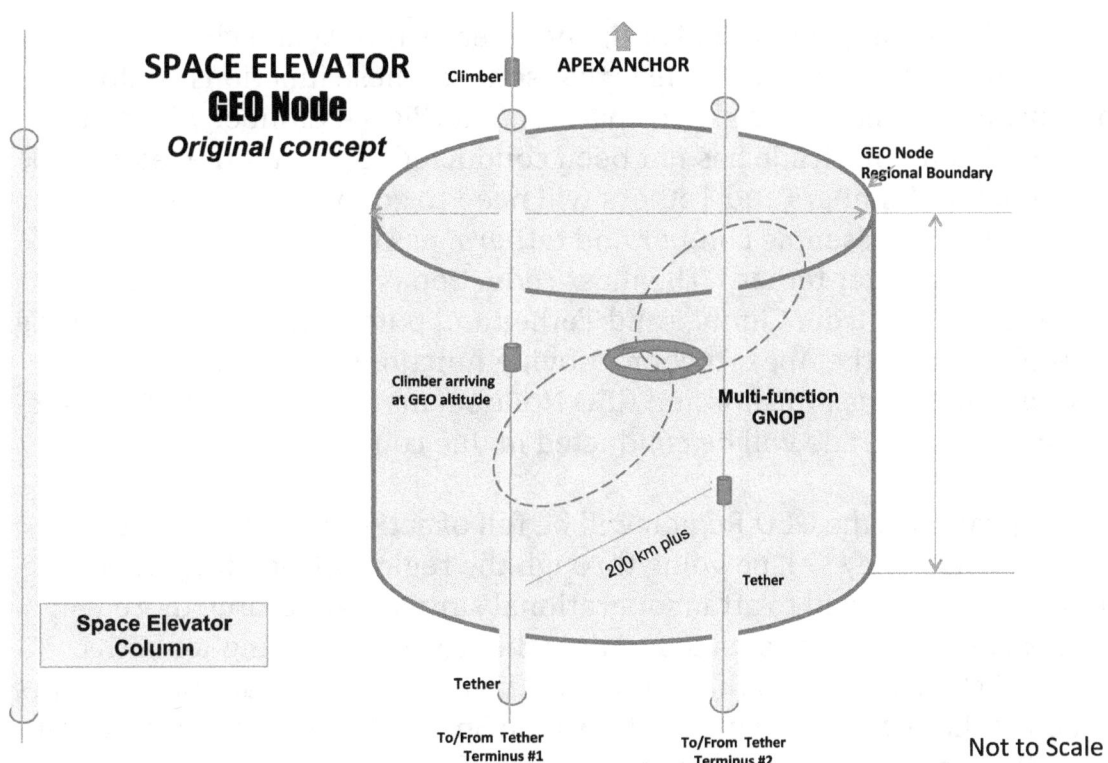

Figure 5 This Operational View was the original GEO Node

4.1.1 Action at the GEO Node; as originally conceived

The only action within the GEO Region was to be the release and/ or capture of payloads to and from Climbers. After a release from a Climber, mission payloads would then do a deployment of appendages, checkout the software, engage the power supply [solar cells to batteries] and go on its own way to its mission location elsewhere in the GEO Belt or perhaps to other orbits; inclined or circular. Of course, there would be action in reverse snaring a satellite for return to Earth. The appendages will be folded or detached and the payload package will be "safed" and folded into the Climber for return to the Earth Port. The satellites themselves are largely responsible for these

31

functions; with some support from a Climber. As we said, "... no active primary function..." for the GEO Node.

With time, it became clear that ISEC needs some "active primary functions" at GEO would be required well before IOC. Active primary functions are needed to test, monitor, and service the entire deployment process. We called them overhead functions. The Pre-IOC deployment and testing process will take years. Further, deployment support functions should be preserved at the GEO Node for post IOC activities.

The original thought process had assigned release, return and check out functions to Climbers. It is now clear that some of these functions could be accomplished by small, specialized spacecraft at GEO. A Climber / Tether / GEO / APEX functions trade has not been conducted. Some on the team hold the opinion that Climbers and Tethers will need to go through a weight reduction effort. Lessening climber and tether mass will lessen stresses on the 100,000 kilometer tether. The mass reduction is best achieved by removing functions from Climber and Tether and putting them elsewhere; e.g. at the GEO Node or the Apex Anchor. Moving functions off Tether and Climber and onto Apex Anchor and GEO Node seems inevitable. Be that as it may, the functions trade will be conducted in due course.

As IOC approaches, the GEO Region will be full of activity from individual tether climbers. They will be going through the region, off-loading and on-loading material needed to attain operational status. In addition, there will be situational awareness systems to monitor the "care and feeding" of both tethers and climbers within the region. This will lead to operational concepts managed at the HQ/POC. Figure 5 shows the Space Elevator Transportation System GEO Node at IOC with a minimum set of responsibilities, yet with great potential for future activities.

4.1.2 Action at GEO at IOC – Transportation and Enterprise

ISEC is already examining what might be in space by the year 2040 and later. The finding was simple: by 2040 – We are not alone. Space Industry members and the government are already actively seeking information about what might be a business basis in 2040; refuel, repair, rescue, and so on. The GEO Node's transportation design considerations must accommodate this future burgeoning space commercial activity. Further, state entities will be in play. Surveillance, safety, and law enforcement have already been "staked out" as roles of national interest; or will be as soon as they become aware of the situation. These considerations led us to forecast a GEO Node Operational View as seen in Figure 6.

The view is of a Space Elevator Transportation System GEO Node, at IOC. Other views of the GEO Node for the Space Elevator Enterprise System will be shown to illustrate the growth.

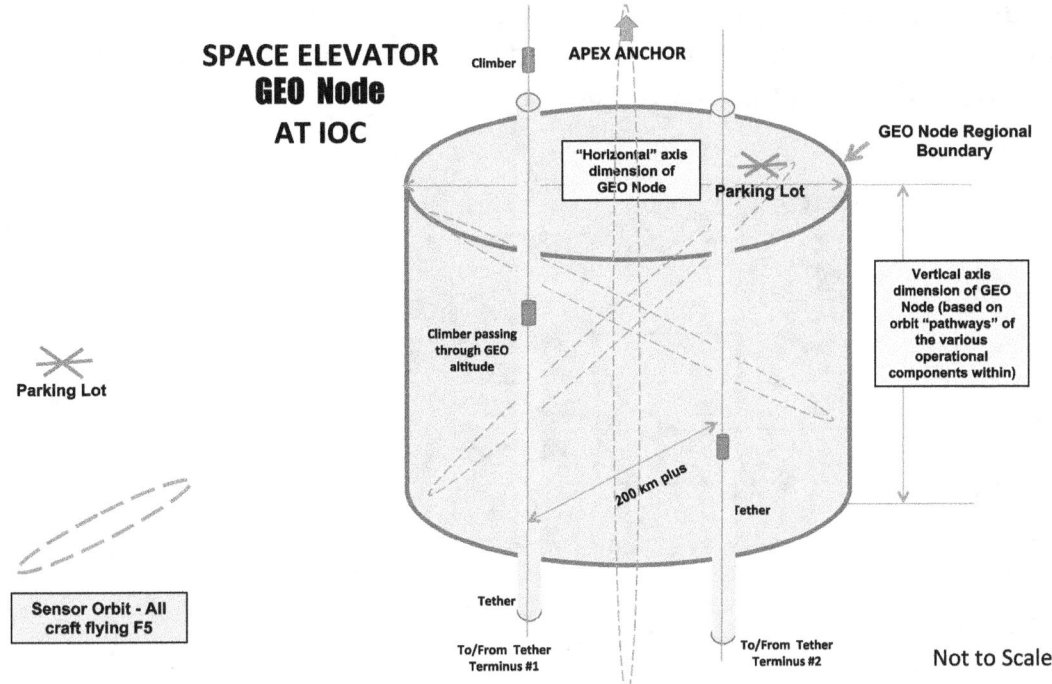

Figure 6 The Operational View shows the GEO Node at Space Elevator Transportation System IOC

4.1.3 Action at GEO at IOC – Two baselines

These considerations led ISEC to commit to the establishment of two baselines; a transportation baseline and an enterprise baseline focused uon supporting business activity in the region. Involvement with nation states will be included in the Enterprise baseline. Given this broader view, ISEC expanded / broadened its GEO Node Transportation view at IOC in Figure 6. It is much more robust than the activities previously portrayed. ISEC foresees a robust surveillance recording activities throughout the Region as well as a parking lot of test support spacecraft and safety operations support craft. Use of that Parking Lot will also be offered to our business clients.

4.1.4 Action at GEO at IOC – Baseline transportation & Sequences.

For IOC to be achieved, a Space Elevator at the GEO Node has two foundational purposes centering on the buildup of the transportation system. The first is to enable the movement of equipment and payloads from the Earth Port to some selected destinations along the tether. The second foundational purpose is to service the sequences – based process to add

33

functions / services. The Space Elevator Enterprise Baseline will be the capture document for all functions that supplement the transportation baseline for delivery of payloads to operational status and stations. In ISEC's Architecture engineering approach, reaching IOC requires successfully passing through the first five sequences and documenting it in Sequence #6 – The Initial Operational Capability Sequence.

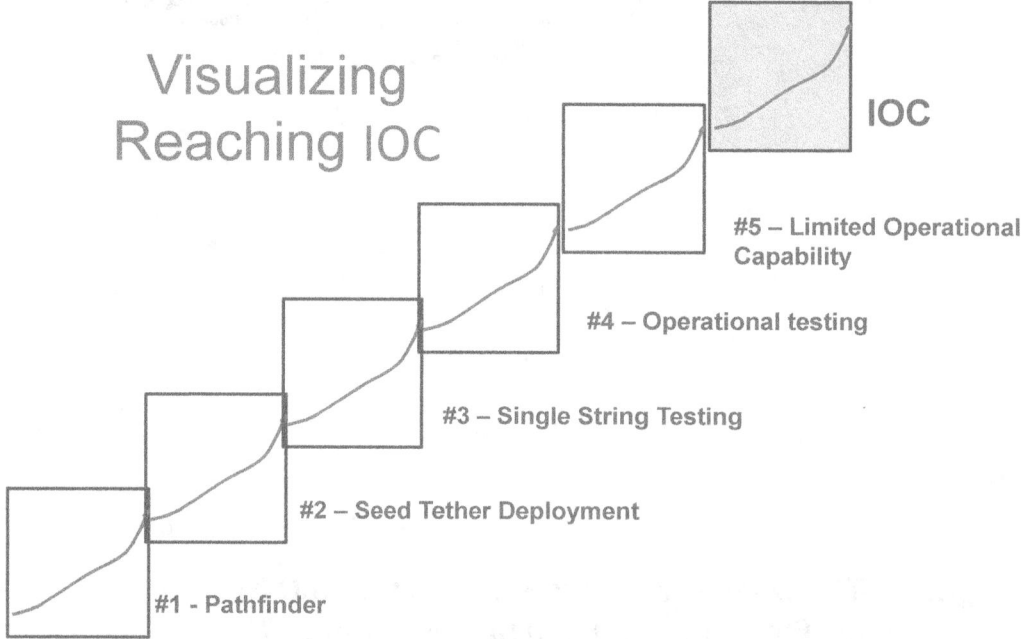

Figure 7 Sequence #6 – The Initial Operational Capability Sequence.

4.1.5   Action at GEO – The Enterprise Post IOC

As soon as IOC is achieved, attention will be turned to adding whatever is needed for our clients' business purposes. The action will follow a dual path. Some of the services and functions will be provided by the Space Elevator Transportation System and some will be provided by an operating entity adjacent to the Space Elevator Transportation System. In the post IOC time frame these adjacent service functions will be added based upon understandings and agreements with Space Elevator clients and customers. In our Architecture engineering approach, adding new functionality requires successfully passing through parallel testing and an on-ramp sequence.

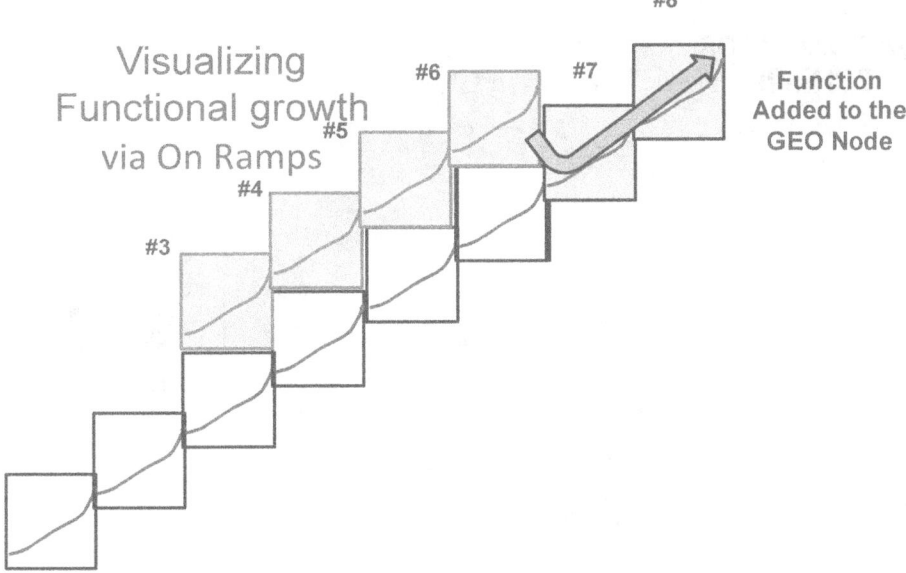

Visualizing Functional growth via On Ramps

#3 #4 #5 #6 #7 #8

Function Added to the GEO Node

Figure 8 To the on-ramp sequence.

The services and overhead functions will be added in an orderly system engineering based fashion. Safety and efficiency will be key standards in this change process. Services and overhead functions will be capabilities at the Space Elevator Enterprise's GEO Node enabled by the Space Elevator Transportation system. The GEO Node in the Enterprise system will be a very busy place. Figures 9 and 10 capture the story.

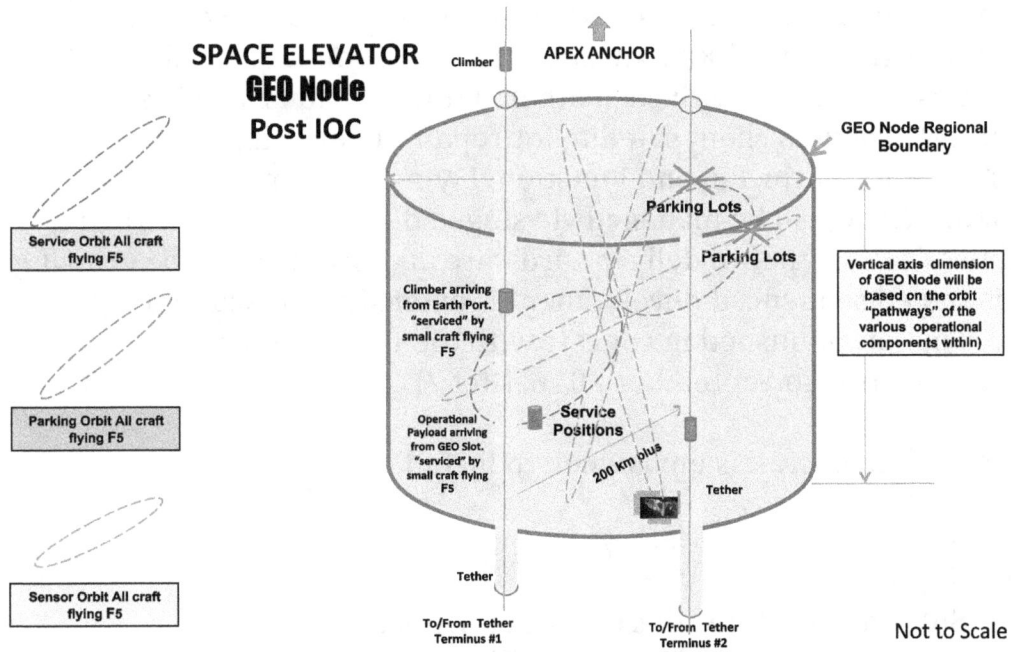

Figure 9 Operational View of the GEO Node within the Space Elevator Enterprise System -- Part One

35

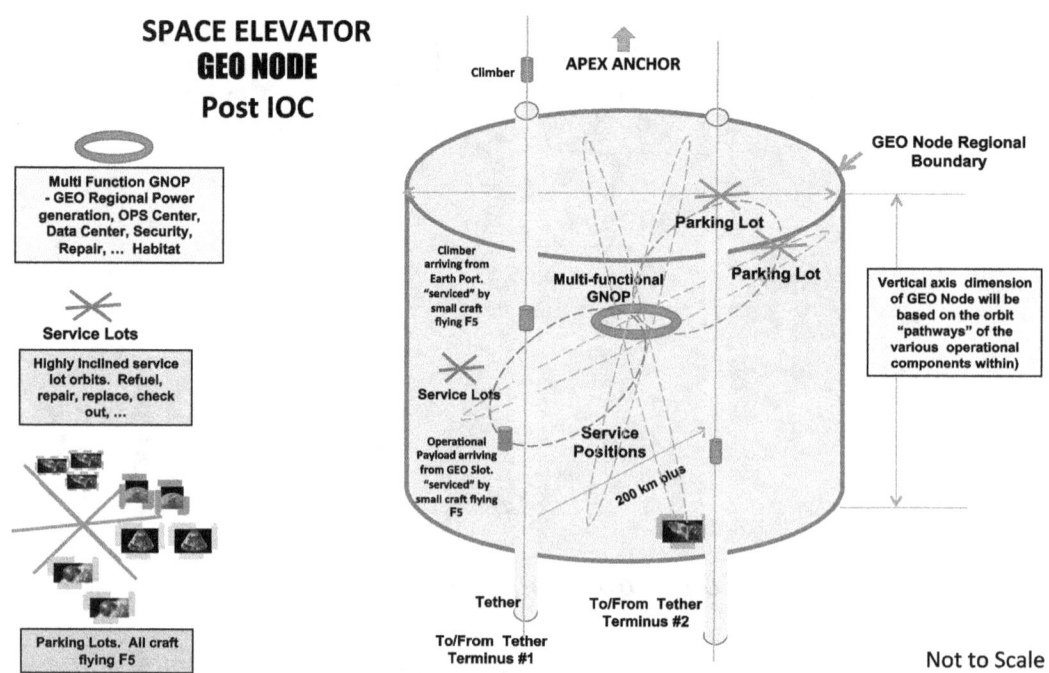

Figure 10 Operational View of the GEO Node within the Space Elevator Enterprise System -- Part Two

4.1.6   Action at GEO – Flying in Formation

ISEC's Modular Construct envisions several services provided by craft gathered in the region. A well-known construct will dominate; small satellites operating in close proximity with each other. ISEC envisions small service craft "gathering" around a client satellite for repair or servicing. We have termed this F5 ➔ for Form, Fit, and Function; Flying in Formation. The growth of a GEO Node will be managed via staged development; leveraging the capabilities of the tether to deliver hardware and products to the growing region. This staged sequenced development is extensively discussed in the IAC-16 D4.3.8 paper; published in concert with the International Astronautical Congress in September 2016. [Ref 7]

An assortment of small craft is envisioned to include:

- Refueling satellites,
- Repair satellites,
- Software system diagnosis satellites,
- Line Replaceable Units satellites,
- Anomaly diagnosis satellites,
- Propulsion satellites ("Space Tug"),

- Power augmentation satellites,
- Communications augmentation satellites, and
- Sensor satellites of several sorts

4.1.7 Action at GEO – GEO Node versus GEO Region activities

The activities of the GEO Node will be taking up room – a lot of room. ISEC expects a GEO Region of over 19 million cubic kilometers! This is based upon a GEO Region estimated to be 200 kilometers tall and 350 kilometers in diameter.

The need for "room at the inn" at GEO starts early; and, some activities will persist for years. Approaching IOC, operational testing and training support needs will involve much of the GEO Region. A variety of assets will be parked at GEO supporting on-orbit deployment testing of all sorts.

Test assets will be deployed to GEO as early as Sequence #3 (Single String Testing) when the latter phases of single string tests are to be conducted. Those support assets will be parked in the GEO Region when Sequences phases are completed. They will be awaiting post IOC modular growth / on-ramp activity. (Every post-IOC modular addition to the Space Elevator will proceed through the steps cited in "Sequences," whether it is for the Transportation system or for the Enterprise. --- see Figure 8)

4.1.8 Action at GEO –GEO Region activities and sub regions

The examination of activities at GEO Node and the room they will claim became an important part of our design consideration. In short, the design considerations must deal with this room thing. We see the Transportation GEO Node 1) supporting transportation operations, 2) enabling the addition of new functions / modules via the Sequences, and 3) storing assets for deferred activities. Thus, various parts of the GEO Region are in different states of operations. We thought it best to define these "parts" as sub-regions, as shown in Figure 11.

37

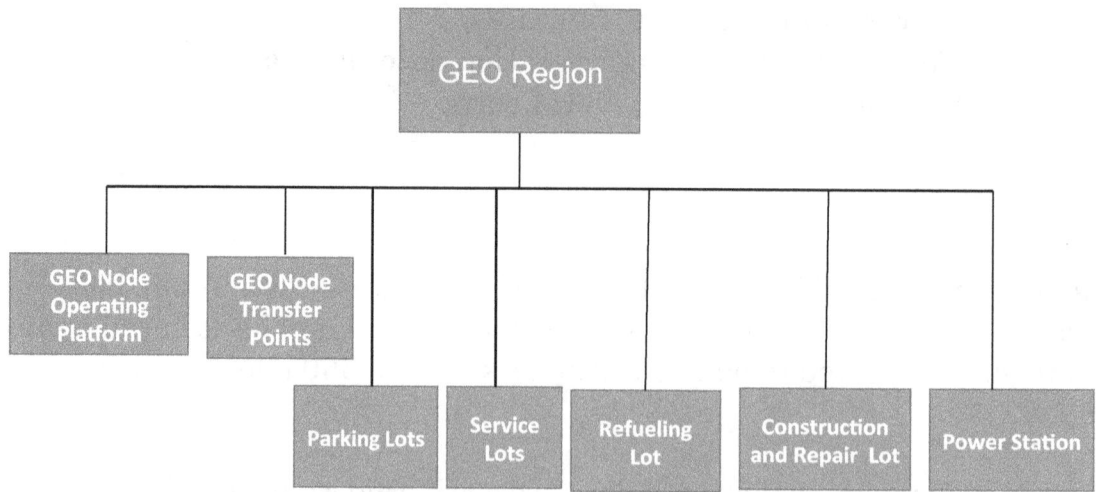

Figure 11 GEO Region / sub-region construct

4.2   GEO Node Study Findings

Soon after the study began, it became clear that the ISEC study group needed to understand the views and dreams of the contributors. The brainstorming session lasted for months and engaged the contributors to determine just what the GEO Node needed to be. The process that ISEC has instigated leverages expertise from its many members and challenges them with a problem. This year-long activity stimulated many ideas and then allowed the team to formulate concepts, ideas, and reasonable conclusions and findings.

4.2.1   Finding – Deployment support at GEO

The GEO Region will grow in a modular, ordered, way. The Modular Construct will be part of its operational style. During Sequence 2 (the Seed Tether Phase) and Sequence 3 (the Single String Testing phase), the GEO region will be the location for the single string tether deployment satellite. From that time, the GEO Region will continue to have one or two tethers flowing through it while each tether is being lengthened to reach the Earth Port's Tether Termini. At first these early phases will have just small tethers being reinforced with tether buildup climbers and then more robust tethers building up to Sequence #5 (Operational Test) and Sequence #6 (Initial Operational Capability). These phases would only have the tethers and, periodically, climbers going through the GEO region. Surveillance of these delicate test and deployment operations will be mandatory. The various support craft will be secured in orderly "Parking Lots." Figure 12 captures this view of the activities at GEO leading up to IOC.

4.2.2 Finding – F5 ➡ Form, Fit, Function, Flying in Formation

We envision that activities at the GEO Node will support achieving IOC with "overhead functions;" 1) asset movement by Space Tug, 2) situational awareness sensing, 3) environmental and test data collection & data pre-processing and, 4) other safety functions. These overhead functions will be executed by specialized spacecraft each uniquely designed to conduct specific operations very near each other and with test & deployment operations craft (e. g. the tether build up climber). The craft will also be parked very close to each other. Proximity operations will be the norm for the GEO Node with both transportation and enterprise delineations. These and overhead operations support functions will be retained post IOC.

4.2.3 Finding – Customer / Client support. Our Modus Operandi

After the Space Elevator IOC, the GEO Node will start growth to become the center point for the Space Elevator's clients and their satellite operations support. Client payloads will arrive at the GEO Node and will be prepared for mission operations, deploy appendages, checkout the software, engage the power supply and go on its way. Clients will also seek to inspect, repair, refuel and augment their operational systems to extend their profitable lifetimes. These activities will be logistically supported by material and Line Replaceable Units (LRU) deliveries by our Space Elevator.

In each case, **F5** craft will gather in formation to execute support operations for both the Space Elevator transportation system and the Space Elevator Enterprise services.

Parking lots at the GEO Node will host many satellite types: refueling, repair, software, LRU, anomaly diagnosis, propulsion ("Space Tug"), power augmentation, communications augmentation, and sensor satellites of several sorts.

The contributors from the 2016 International Space Elevator Conference workshop saw a variety of commercial activities after IOC, such as: factories, research, interplanetary travel, satellite servicing, power generation, tourism, mission satellite orbital placement [space tug], and much more.

Last year's study report, "Design Considerations of a Space Elevator Earth Port," established a concept that the Earth Port was a complex association of many tasks and missions [Ref 6]. The Earth Port's iconic regional portrayal established the relationship between the Earth Port, the Apex Anchor and the GEO Node. This intimate, connected relationship between the three regional elements of the transportation system makes our physics work.

The Earth Port, as shown in Figure 12, will:
- serve as a mechanical and dynamical termination of the space elevator tether, providing reel-in/reel-out capability and position management
- serve as a port for receiving and sending Ocean-going Vessels (OGVs) with climbers, payloads, supplies and personnel
- provide landing pads for helicopters
- serve as a facility for attaching and detaching payloads
- provide tether climber power for the 40 km above the Floating Operations Platform (FOP)
- provide food and accommodation for crew members as well as power, desalinization, waste management and other such support.

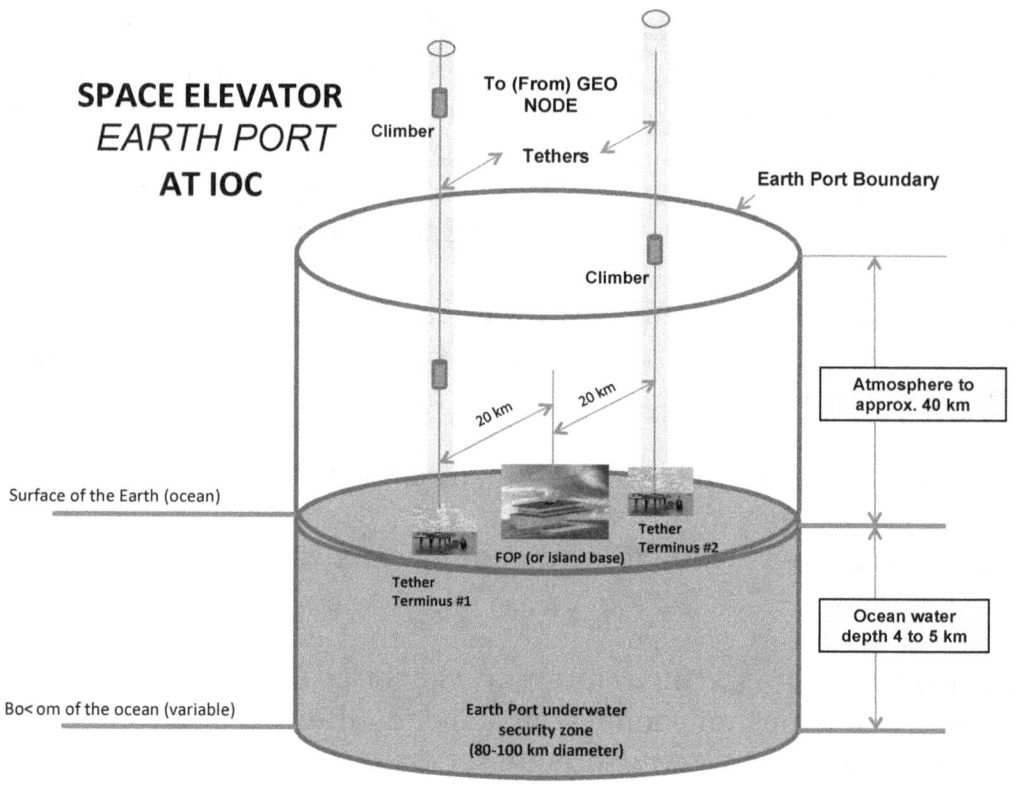

Figure 12 Earth Port Elements

During this study's brainstorming sessions, ISEC recognized that each segment was an extension of another. Early on, the GEO Node study team saw that the number of Tether Transfer points at GEO should match the number of Tether Termini at the Earth Port (aka the "Mirror Image approach"). What goes on in one segment affects other segments. This parallel vision is shown in the image on the cover of this study report.

4.2.5 Finding - The Essential Space Elevator IOC

At the end of the brainstorming period, the team began to discuss and understand the "essential" Space Elevator. The team knew that only a minimal number of functions should be part of any system's IOC. So, after some discussion, it was determined that the definition of IOC for the Space Elevator would be based upon the functions needed for the Space Elevator Transportation System.

ISEC's Architecture and Roadmaps document [Ref 4] does not offer an extensive definition of the Space Elevator IOC. However, that 2014 report correctly portrays pathways and hurdles to achieve segment IOCs, and identifies the key efforts to bring the segments together into an operating transportation system. That thesis and theme remain correct. ISEC has only updated the definition of IOC:

Definition of Initial Operational Condition

The Space Elevator Transportation System is comprised of one Earth Port with two tether termini, an Apex Anchor supporting two 100,000 km Tethers, 14 Tether Climbers, and a single Headquarters and Primary Operations Center. The GEO Node supports the Space Elevator Transportation System with a range of "overhead' functions; e. g. test, safety, and support.

4.3   Functional Baseline of the GEO Node

The new ISEC finding is that there are two baselines that must be preserved. The position that there are no active, primary functions needed at the GEO Node for the Space Elevator Transportation system remains. There are plenty of essential overhead functions needed:

1) Asset movement by Space Tug,
2) Situational awareness sensing,
3) Environmental and test data collection & data pre-processing
4) Safety and surety functions

4.3.1 GEO Node Transportation Functions

GEO Node functions, before IOC, are to test, monitor, and service the entire deployment process. The pre-IOC deployment and testing process will take years. These same deployment support functions will be preserved at the GEO Node for the post IOC On-ramp activities.

As it approaches IOC, the GEO Region will be full of activity from individual tether climbers. They will be going through the region – off loading, on loading – and strengthening/repairing the tether. In addition, there will be situational awareness to ensure "care and feeding" of the tethers and climbers close to the region. All these will lead to operational concepts being handled at the HQ/POC. Figure 13 shows the view that the GEO Region at IOC will have minimum responsibilities with great potential for the future.

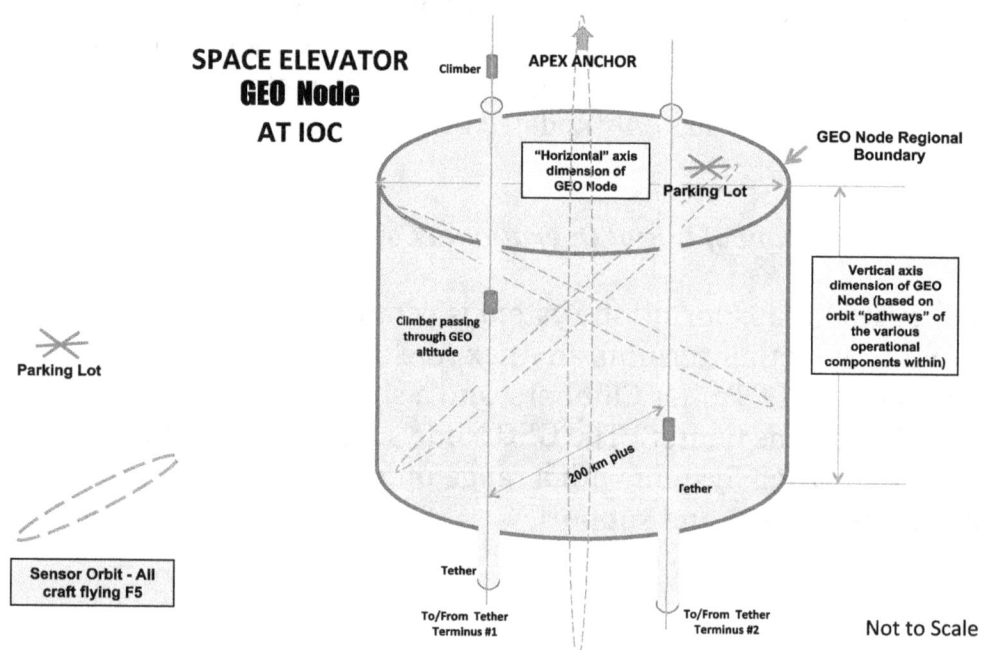

Figure 13 GEO Node at IOC

The family of functions identified in the GEO Node study's brainstorming phase will be preserved and evaluated for inclusion in the GEO Node Enterprise baseline. These Enterprise functions most likely will be services preferred by customers or by investors for business-based improvements.

It is implicit in this situation that these latter functions will be inserted, in a modular growth pattern, into the Space Elevator Architecture after passing through the engineering test and validation phases. The Modular Growth process is still being defined; but, it will clearly meet or exceed the standards cited in Sequences.

4.3.2 The GEO Node Buildup

Many of the possible post-IOC functions that emerged in the brainstorming process were "service" functions--- functions that a customer wants (or might want) to better execute the mission of the clients' payloads --- not the mission of a Space Elevator as a transportation system. The GEO Node will transition from the role of transportation assets to one of business enabler within a very valuable region. Addressing the 500 GEO satellites operating today, it is apparent that there are plenty of businesses that will spring up when access to the location is low cost, routine and safe. Not only will the new missions be parallel with all the ones in existence, but many missions that are not even conceived of today will be mainline then.

A process will be established for the introduction of new functions and services. It is now clear that the services/functions brought on will be of two types: Transportation System improvements and Enterprise support services driven by client needs. It is also evident that the distinction will not always be obvious; so, some "jointly owned" considerations will be part of the growth process. For the sake of our dual baseline documentation, these jointly owned functions will be assigned to one baseline or another at the time.

4.3.3 GEO Node change process – orbital bypass surgery

We must conceive how these functionalities can meld into the operational architecture which will be in place at that time. It is clear the architecture must operate through the period in which new functions are being added; AND, that portions of the architecture will change their manner of operation. Notionally, we foresee that changed portion will be bypassed in some manner until the changeover is completed.

4.3.4 GEO Node Enterprise Functions

The following are likely or possible Functions at the GEO Node

- Communications: The GEO Node communications functions will largely relate to the transactional communications within the region to support F5 modular operations and the timely reporting of situational awareness concerns and operational surety. The GEO Region Operations Control Center will need distinct / unique embedded auto-communications functions to support collision avoidance, debris detection, intrusion, environmental event detection and full situational awareness.

- Maintenance: The modular design standard will emphasize Line Replaceable Units enabled by robotic replacement schemes. New or improved LRU modules must be cited as acceptable by appropriate engineering processes.

- Power Generation: All sorts of power need to be generated to execute GEO Node functions and services. It may be that a mix of solar and space nuclear is needed. Energy Multiplier Modules' (EM²) nuclear elements are developing apace with Space Elevator needs.

- Propulsion: All objects within the GEO Node must actively maintain orbital position, avoid collisions, and manipulate orbital position to execute or receive function / service. A variety of propulsion types are expected to be used and maintained

- Payload Processing: Provide services as per customers' request

- Payload Anomaly Diagnosis: A special case pf payload processing provided as per customer request

- Local Operations Controls (Center): A sub element of HQ/POC with some local autonomy for collision avoidance, debris detection, space object intrusion, environmental event detection and GEO Region situational awareness.

- The GEO Node will have some responsibility for controlling tether dynamics (i. e. Reel-In / Reel-Out) in collaboration with the Earth Port, APEX Anchor and HQ Operations

- Refueling: Refuel GEO Node assets and payloads as part of servicing. Safety, surety, and environmental restrictions will be considered.

- Payload Transfer: A series of handling functions that enable payload/cargo movement from and to interim and final destinations.

- Security & Surveillance: The GEO Node will provide a range of surveillance sensing to support client's and GEO Node operations.

- Safety: The GEO Node will be a safe operational environment

Habitat: Human Habitation Operations is not provided. GEO Node Operations will, for now, be robotic.

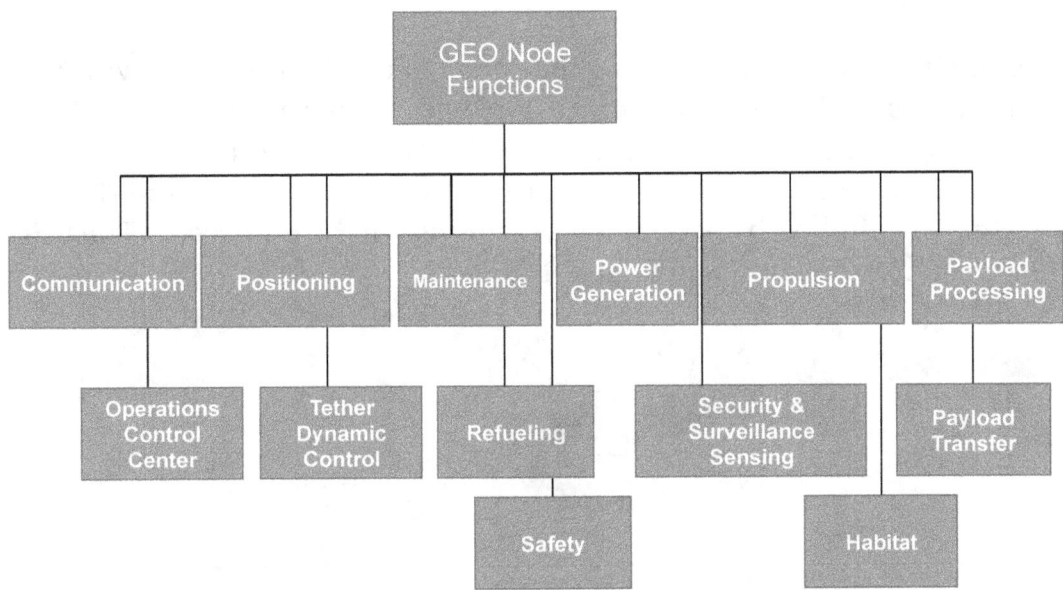

Figure 14 GEO Node Function supporting the Space Elevator Transportation and Enterprise Systems

4.3.5 GEO Node growth standards

Bringing service functions to the Space Elevator that are service functions is going to be an interesting series of value judgments: should it be done in-house or be outsourced? One would then wonder where these value judgments would get the empirical information (e. g. test or simulation information) needed. We foresee that the on-ramp testing and mission evaluation would be conducted somewhere along the length of the Tether – perhaps at the GEO Node.

The "foundation for future growth" is a function held by the GEO Node for the entire Elevator. We also saw the GEO Region as the place that does certain string tests, certain operational tests, and some possible limited operational capability testing for functions destined to be on-ramped into the Space Elevator.

At the heart of our GEO Node vision is a consensus that we are dealing with a modular standard. We embrace this standard for two approaches: for Architecture growth and operations.   The core of this standard is known as F5. F5 stands for Form, Fit, and Function; Flying in Formation. Pieces within the GEO Region will gather as F5 and interact using transactional communications. Connect without contact.

The preceding section spoke of adding functions to the Architecture with on-ramps and bypasses. The ordered traffic flow within the GEO Region will be from sub region to sub region within the GEO Region. Operational craft within each sub region will be a mixture of Transportation assets and Client assets. Defined orbital paths between sub regions will be enforced.   The GEO Node Region concept is shown in Figure 15.

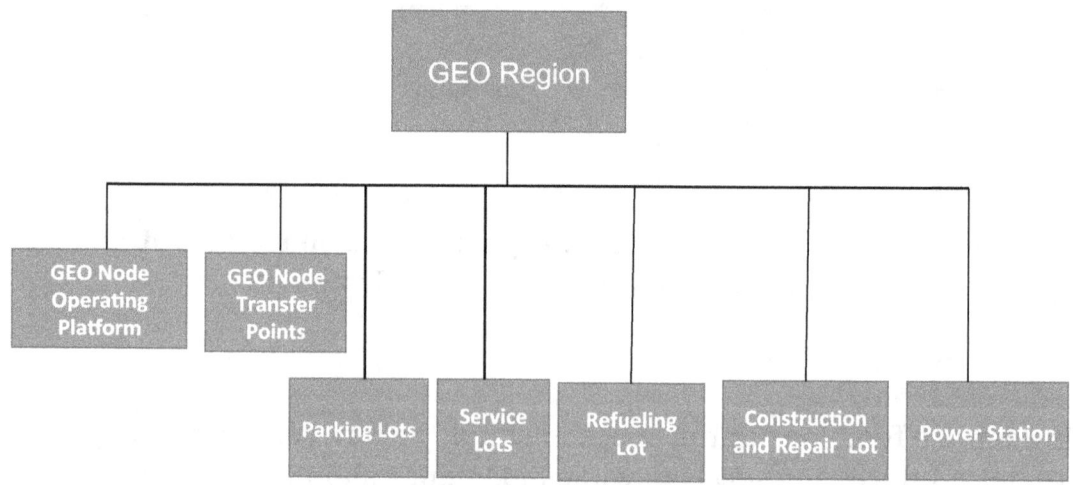

Figure 15 Region / Sub-Region Construct

Possible GEO Node Sub-Regions are:

- Parking lots for service craft, customer payloads
- Service lots for service on customer payloads including refueling
- Construction and repair lots
- Power Generation facility, and
- Operations & surveillance station including data processing and communications

Many of the contributors wrote and commented eloquently and earnestly about what they thought a GEO Node would need to be. The early months were an explosion of ideas and worries. One day someone wrote of factories and hotels vigorously operating in or near the GEO Region; and then, that same person would worry about the tether whipping around dangerously without control. The contributors mellowed and the resulting vision and compilation of work products is an astounding array of vision and practicality.

At first the FOC GEO Node was portrayed as a "Space Station" at GEO with adjacent transfer points for arriving and departing climbers bringing and taking valuable cargo. The notions matured and after time the "Space Station" was transformed from an assembled structure to a gathered group of capabilities flying in close formation. The transfer points became simply identified locations at which GEO Node based service craft greeted arriving climbers processing the payload for continued travel.

Between arriving climbers, the service craft await the next service call in parking locations at slightly inclined controlled orbits. The GNOP (GEO Node Operations Platform) located near the orbital centroid of the GEO Region is also in a slightly inclined orbit. All objects within the GEO Node will have their own orbit maintenance propulsion or be 'moved" by a provided space tug. In some cases, a combination of propulsions might be called for. Any craft or object within the GEO Region must have sufficient propulsion to meet safety and surety standards.

A pair of figures shown next [Fig 16 a & b] how an evolved GEO Node might look.

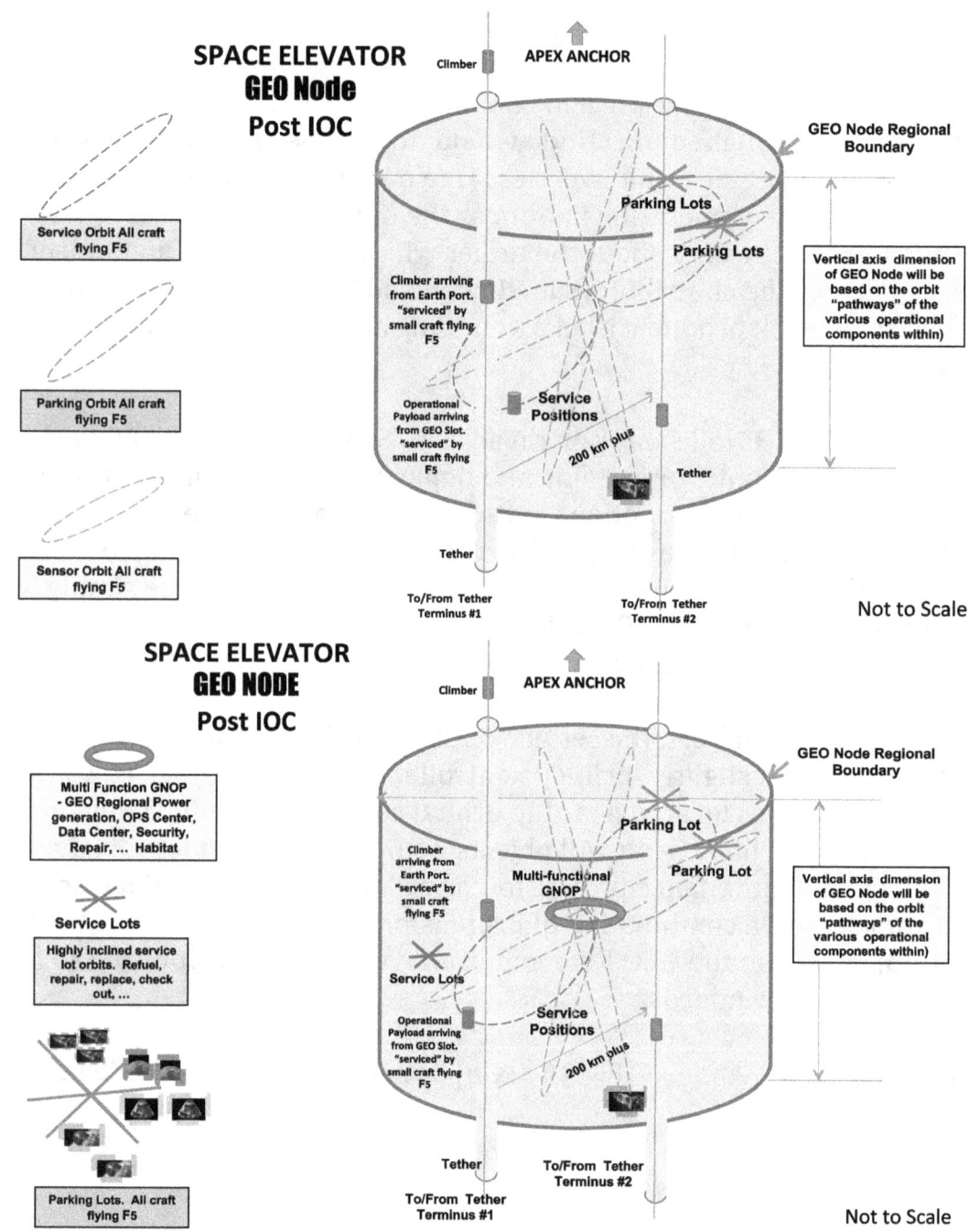

Figure 16a and 16b Evolved GEO Node Pictorials (image By Michael Fitzgerald)

4.4 Conclusions and Recommendations

4.4.1 Conclusions

The GEO Node is foundational to the future, successful, Space Elevator Enterprise. It is the primary mission enabling interface between the Space Elevator Transportation System and the Space Elevator Enterprise.

The GEO Node is expected to become the centerpiece of a Galactic Harbour that provides services such as repair/assembly, refueling climbers, loading and offloading supplies, servicing tugs and many other things for a myriad of space-faring customers.

Today's technology should suffice to understand the needs of the Space Elevator and its customers. However, future technological innovations will indeed enhance capabilities in space and especially at the GEO Node complex.

While evaluating and developing the Apex Anchor and GEO Node concepts, one must be cognizant of Earth Port design characteristics. Indeed, there are several parallels within this tremendous space elevator transportation infrastructure between the Earth Port and both the GEO Node and Apex Anchor. One of ISEC's favorite images shows this in the cover image.

4.4.2 Recommendations

Continue to identify and catalog potential functions and services for the GEO Node with an eye to enhancing the overall vision for Space Elevator investors.

References:

[Ref 4] Fitzgerald, M, R. Penny, P. Swan, C. Swan, Space Elevator Architectures and Roadmaps, ISEC Study Report, lulu.com, 2015

[Ref 6] Hall, Vern, R. Penny, P. Glaskowsky, S. Schaeffer, Design Considerations for Space Elevator Earth Port, ISEC Study Report, www.lulu.com, 2016

[Ref 7] Swan, Peter, Michael Fitzgerald, "Space Elevator Development Sequence," International Astronautical Congress, IAC-16-D4.3.8, Guadalaraja, 2016.

5 Study Summary

Preliminary Portrait of ISEC's Vision

Figure 16 did not strike like a bolt out of the blue. Rather, it is the first total capture of what we all have been seeing. It adds perspective. We all embrace it. This figure illustrates the three regions ranging from Earth Port, through the GEO Node to the Apex Anchor.

Figure 17, Space Elevator Regions

The Earth Port is essential. The unique orbital physics at the GEO Node's geosynchronous altitude offers "gathering" as a real modus operandi for servicing, commerce, and production. Finally, there is no question that the APEX - which includes the gates to the Moon and Mars – offers travel to destinations that we have only imagined. All of this because we dare to think that we can build the Space Elevator. [Note: The horizontal distances are not to scale as the Earth Port diameter is roughly 40 kms, the GEO Region about 265 kms and the Apex Anchor Region about 670 kms in diameter.]

Conclusions

- The deployment and continued stability of the tether are the primary function of the Apex Anchor until IOC. This translates to a) a reel in/reel out (or climb up/climb down) capability, b) the capability to fire thrusters (magnitude and direction) as directed by HQ/POC, and c) support to customers who leverage the strength of the end point of this space transportation infrastructure.

- The basic mass buildup for the Apex Anchor will be from spent climbers and derelict GEO satellites initially.

- The GEO Node is expected to become the centerpiece of a Space Port that provides "overhead" services such as repair/assembly, refueling climbers, loading and offloading supplies, servicing tugs and many other things to a myriad of customers.

- Today's technology should suffice to understand the needs of the Space Elevator and its customers. However, the future technological capabilities will indeed enhance capabilities in space and especially at the GEO Node complex.

- While one is evaluating and developing the Apex Anchor and GEO Node concepts, one must be cognizant of the Earth Port design characteristics. Indeed, there are several parallels within this tremendous space elevator transportation infrastructure between the Earth Port and both the GEO Node and Apex Anchor.

Recommendations

- Incorporate the new definitions into the Lexicon for space elevator terminology.

- Stimulate further research in development of the Apex Anchor, especially with respect to the accumulation of the masses for both the tether growth and Apex Anchor stability needs.

- Continue to identify and catalog potential functions and services for the GEO Node Region with an eye to enhancing overall Space Elevator investments.

Who We Are

The International Space Elevator Consortium (ISEC) is composed of individuals and organizations from around the world who share a vision of humanity in space.

Our Vision

A world with inexpensive, safe, routine, and efficient access to space for the benefit of all mankind.

Our Mission

The ISEC promotes the development, construction and operation of a space elevator infrastructure as a revolutionary and efficient way to space for all humanity.

What We Do

- Provide technical leadership promoting development, construction, and operation of space elevator infrastructures.
- Become the "go to" organization for all things space elevator.
- Energize and stimulate the public and the space community to support a space elevator for low cost access to space.
- Stimulate science, technology, engineering, and mathematics (STEM) educational activities while supporting educational gatherings, meetings, workshops, classes, and other similar events to carry out this mission.

A Brief History of ISEC

The idea for an organization like ISEC had been discussed for years, but it wasn't until the Space Elevator Conference in Redmond, Washington, in July of 2008, that things became serious. Interest and enthusiasm for a space elevator had reached an all-time peak and, with Space Elevator conferences upcoming in both Europe and Japan, it was felt that this was the time to formalize an international organization. An initial set of directors and officers were elected and they immediately began the difficult task of unifying the disparate efforts of space elevator supporters worldwide.

ISEC's first Strategic Plan was adopted in January of 2010 and it is now the driving force behind ISEC's efforts. This Strategic Plan calls for adopting a yearly theme to focus ISEC activities. (For 2010, the theme was "Space Elevator Survivability -- Space Debris Mitigation.") In 2010, ISEC also announced the first annual Artsutanov and Pearson prizes to be awarded for "exceptional papers that advance our understanding of the Space Elevator." Because of our common goals and hopes for the future of mankind off--planet, ISEC became an Affiliate of the National Space Society in August of 2013.

Our Approach

ISEC's activities are pushing the concept of space elevators forward. These cross all disciplines and encourage people from around the world to participate. The following activities are being accomplished in parallel:

- CLIMB – This annual peer reviewed journal invites and evaluates papers and presents them in an annual publication with the purpose of explaining technical advances to the public. The first issue of CLIMB was dedicated to Mr. Yuri Artsutanov (a co-inventor of the space elevator concept); and, the second issue was dedicated to Mr. Jerome Pearson (another co--inventor). CLIMB is scheduled for publication each July. They can be downloaded at www.isec.org.
- Yearly conference – International space elevator conferences were initiated by Dr. Brad Edwards in the Seattle area in 2002. Follow--on conferences were in Santa Fe (2003), Washington DC (2004), Albuquerque (2005/6 –smaller sessions), and Seattle (2008 to the present). Each of these conferences had multiple discussions across the whole arena of space elevators with remarkable concepts and presentations. Recent conferences have been sponsored by Microsoft, the Seattle Museum of Flight, the Space Elevator Blog, the Leeward Space Foundation, and ISEC.
- Yearlong technical studies – ISEC sponsors research into a focused topic each year to ensure progress in a discipline within the space elevator project. The first such study was conducted in 2010 to evaluate the threat of space debris. The second study, and resulting report, focused on space elevator operations. The 2013 study focused upon tether climber designs. The 2014 topic is Space Elevator Architectures and Roadmaps. There is one topic chosen for 2015; Earth Port Design Considerations. The products from these studies are reports that are published to document progress in the development of space elevators. They can be downloaded at www.isec.org.
- International cooperation – ISEC supports many activities around the globe to ensure that space elevators keep progressing towards a developmental program. International activities include coordinating with the two other major societies focusing on space elevators: the Japanese Space Elevator Association and EuroSpaceward. In addition, ISEC supports symposia and presentations at the International Academy of Astronautics and the International Astronautical Federation Congress each year.

- Competitions – ISEC has a history of actively supporting competitions that push technologies in the area of space elevators. The initial activities were centered on NASA's Centennial Challenges called "Elevator: 2010." Inside this were two specific challenges: Tether Challenge and Beam Power Challenge. The highlight came when Laser Motive won $900,000 in 2009, as they reached one kilometer in altitude racing other teams up a tether suspended from a helicopter. There were also multiple competitions where different strengths of materials were tested going for a NASA prize – with no winners. In addition, ISEC supports the educational efforts of various organizations, such as the LEGO space elevator climb competition at our Seattle conference. Competitions have also been conducted in both Japan and Europe.
- Publications – ISEC publishes a monthly e--Newsletter, its yearly study reports and an annual technical journal [CLIMB] to help spread information about space elevators.   In addition, there is a magazine filled with space elevator literature called Via Ad Astra.
- Reference material – ISEC is building a Space Elevator Library, including a reference database of Space Elevator related papers and publications.
- Outreach – People need to be made aware of the idea of a space elevator. Our outreach activity is responsible for providing the blueprint to reach societal, governmental, educational, and media institutions and expose them to the benefits of space elevators. ISEC members are readily available to speak at conferences and other public events in support of the space elevator. In addition to our monthly e--Newsletter, we are also on Facebook, Linked In, and Twitter.
- Legal – The space elevator is going to break new legal ground. Existing space treaties may need to be amended. New treaties may be needed. International cooperation must be sought. Insurability will be a requirement. Legal activities encompass the legal environment of a space elevator -- international maritime, air, and space law. Also, there will be interest within intellectual property, liability, and commerce law. Starting work on the legal foundation well in advance will result in a more rational product.
- History Committee – ISEC supports a small group of volunteers to document the history of space elevators. The committee's purpose is to provide insight into the progress being achieved currently and over the last century.
- Research Committee – ISEC is gathering the insight of researchers from around the world with respect to the future of space elevators. As scientific papers, reports and books are published, the research committee is pulling together this relative progress to assist academia and industry to

progress towards an operational space elevator infrastructure. For more visit http://isec.org/index.php/about-isec/isec-research-committee

ISEC is a traditional not-for-profit 501 (c) (3) organization with a board of directors and four officers: President, Vice President, Treasurer, and Secretary. In addition, ISEC is closely associated with the conference preparation team and other volunteer members.   Address: ISEC, PMB 204, - 9272 Jeronimo Rd Ste 107A, Irvine, Ca 92618-1978 inbox@isec.org / www.isec.org

Appendix B   Acronyms and Lexicon

CNT	Carbon Nano Tube
CO	Commanding Officer
FOC	Full Operational Capability
FOP	Floating Operations Platform
GEO	Geosynchronous Earth Orbit
HQ/POC	Headquarters Primary Operations Center
IAA	International Academy of Astronautics
IOC	Initial Operational Capability
ISEC	International Space Elevator Consortium
kg	kilogram
LOA	Length Overall
MT	metric ton
MW	megawatt
NASA	National Aeronautics and Space Administration
NCEP	National Centers for Environmental Prediction
NOAA	National Oceanographic and Atmospheric Administration
OCC	Operations Control Center
OGV	Ocean Going Vehicle
PFOP	Primary Floating Operations Platform
STOL	Short Takeoff & Landing
VTOL	Vertical Takeoff & Landing
XO	Executive Officer

IAA study group #3-24 met in Seattle in August of 2015. The team agreed to use, as much as possible, consistent terminology for this report. Below are those terms shown in the figure. . This general list of terminology is shown in the next table: The agreed upon terms should be:

Apex Anchor Node	LEO Gate	Earth Port
Mars Gate	Lunar Gravity Center	- Earth Terminus
Moon Gate	Mars Gravity Center	- Floating Operations Platform
GEO Node	Tether Climbers	Headquarters and Primary Operations Center

Figure - Space Elevator System Lexicon Example

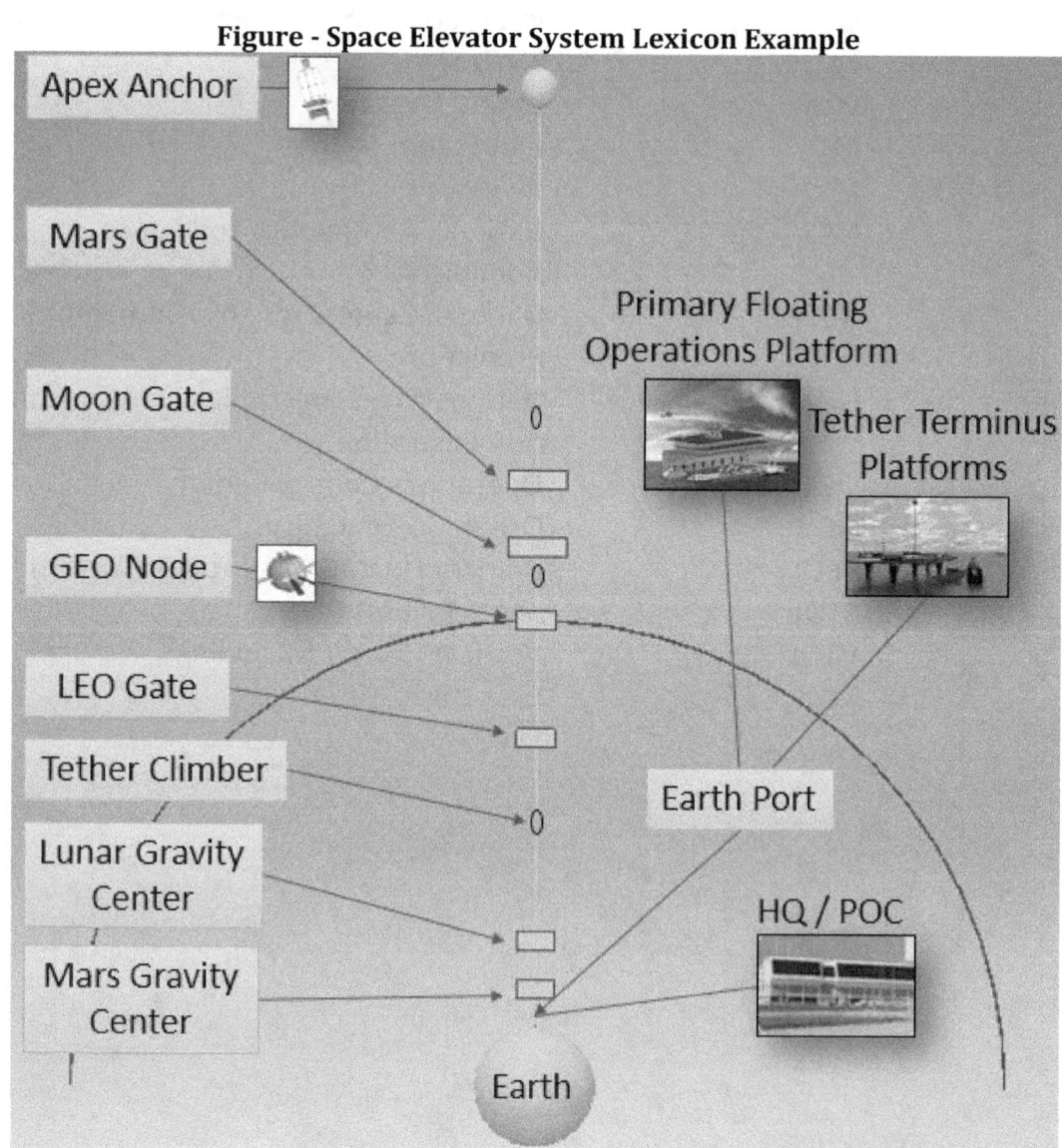

Suggested Terminology

Terminology	Explanation
Apex Anchor	A complex of activity is located at the end of the Space Elevator providing counterweight stability for the space elevator as a large end mass. Attached at the end of the tether will be a complex of Apex Anchor elements such as; reel-in/reel-out capability, thrusters to maintain stability, command and control elements, etc.. [Note: nothing stays at that altitude unless attached to a tether]
Apex Anchor Region	The region around the Apex Anchor is defined by the amount of motion expected at the full extension of the tether. The region is the volume swept out by the end of the tether during normal operations. When two or more space elevators are operating together, the region spreads to the volume between.
Boron-Nitride Nanotube (BNNT)	High Tensile Strength material under development
Capability On Ramps leading to FOC	Time after IOC when new businesses / capabilities are added to system [7th sequence step]
Carbon Nanotube (CNT)	High Tensile Strength material under development
Climbers [Tether Climbers]	Vehicle able to climb or lower itself on the tether
Deployment	Releasing the tether from the GEO construction up and or down during the initial phase of construction
Earth Anchor	Earth Terminus for space elevator
Earth Port	A complex located at the Earth terminus of the tether to support its functions. These mission elements are spread out within the Earth Port Region. When there are two or more termini of tethers, the Earth Port reaches across the region and is considered one Earth Port.
Earth Port Region	The volumetric region around each Earth Port to include a space elevator column for each tether and the space between multiple tethers when they operate together. The Earth Port Region will include the vertical volume through the atmosphere up to where the space elevator tether climbers start operations in the vacuum and down to the ocean floor.
Final (Full) Operational Capability	Design for full capability of the space elevator [8th sequence step]
Floating Operations Platform	The Op's Center for the activities at the Marine Node or Earth Terminus [Earth Port]
GEO Node	The complex of Space Elevator activities positioned in the Space Elevator GEO Region of the Geosynchronous belt [36,000 kms altitude]; directly above the Earth Port. There will be several sub nodes; one for each tether, one for a central main operating platform, one for each "parking lot", and others.
GEO Region	Encompasses all volume swept out by the tether around the Geosynchronous altitude, as well as the orbits of the various support and service spacecraft "assigned" to the GEO Region. When two or more space elevators are operating together, the region includes each and the volume between elevators.
Headquarters and Primary Operations Center [HQ/POC]	Location for the Operations and Business Centers – probably other than at Marine Node
Initial Operational Capability	A term to describe the time when the space elevator is prepared to operate for commercial profit – robotically [6th sequence step]
International Academy of Astronautics (IAA)	International Association focusing upon space capabilities with approximately 1,000 elected members.
International Space Elevator Consortium (ISEC)	Association whose vision is: A world with inexpensive, safe, routine, and efficient access to space for the benefit of all mankind.
Japanese Space Elevator Association	JSEA handles all the space elevator activities for universities and STEM activities. Also handles the global aspects of space elevators.
Japanese Space Agency (JAXA)	Japanese government organization responsible for space systems and space operations.
Length Overall	Full length of the space elevator, est. from 96,000 to 100,000 km
LEO Gate	Elliptical release point for LEO – roughly 24,000 kms altitude
Limited Operational Capability	Early utilization of a "starter" tether in parallel with testing and further development [5th sequence step]
Lunar Gate (Moon Gate)	Release Point towards Moon – roughly 47,000 kms altitude
Lunar Gravity Center	Point on Tether with Lunar gravity similarity – 8,900 kms altitude
Marine Node (Earth Port)	Earth Terminus for space elevator
Mars Gate	Release Point to Mars – roughly 57,000 kms altitude

Mars Gravity Center	Point on Tether with Mars gravity similarity – 3,900 kms altitude
Ocean Going Vehicle (OGV)	Vehicle able to travel over the open ocean
Operational Testing	Key developmental phase when checking out capability [4th sequence step]
Pathfinder	In-orbit testing of space elevator with as many segments represented as possible [1st sequence step]
Primary Operations Center	Center of all activities for the space elevator. Could be distributed or centralized.
Seed Tether [Ribbon]	The initial tether lowered from GEO altitude which would then be built up to become the space elevator tether [2nd sequence step]
Single String Testing	Single string tests are tests conducted of a selected set of Space Elevator functions; aligned and operating. In early forms, single string testing could be an end-to-end simulation of a segment. Later, hardware is inserted in the string to add realism. Testing the initial tether after deployment would be a key single string test.
Space Elevator Column	The volume swept out during normal operations starting at the Earth Port [a circular area within which it operates] and extending through the GEO Region up to the Apex Region. This column of space will be monitored, restricted, and coordinated with all who wish to transverse the volume.
Tether	100,000 km long woven ribbon of space elevator with sufficient strength to weight ratio to enable an elevator [CNT material probably]
Tether Climbers	Vehicle able to climb or lower itself on the tether, as well as releasing or capturing satellites for transportation or orbital insertion.

Appendix C   Brainstorming Session Minutes

Question: How large does the Apex Anchor need to be?
Two Million kg, approximately, but if it is larger, you can shorten the tether. Conversely, if it is smaller, it can be farther out. There is a point in space at which an anchor is no longer needed, but then, there would be no station for tension control. The chosen length account for optimization.

Q: How many G's (Gravitational Forces) do you need to move the space elevator laterally?

A: There is a paper addressing this very topic by Jerome Pearson from 1975. In addition, there will be thrusters at multiple locations along the tether, all the way down to the earth port. Peter Robinson has addressed this issue in his Metrology System paper.

Statement: The tether will need tension sensors at various points along it.

S: The Apex Anchor cannot be too near the moon because of the influence of lunar forces.

This brought up the topic of variables. Some are known, others are predictable, while some are yet unknown. We will need a model that accounts for these variables.

Q: How much energy is needed to launch from the Apex Anchor to the LaGrange point of the moon?

A: Almost nothing; more a calculation of direction upon release.

Q: Would the Apex Anchor be a manned station?

A: The Obayashi Corporation projects having 5 humans in their model. Ours could be fully automated or manned.

S: The initial purpose of the tether would be just to put "stuff" into orbit, but eventually, we will want to bring "stuff" down the tether to earth.

Q: Why use the tether to bring stuff down when we can just fly "stuff" down to orbit and then allow it to return to earth from there? Wouldn't it be more difficult to line up with the Anchor as a single point in space?

A: We still need fuel to slow the descent and vehicles to resist burning up in the atmosphere. The elevator would allow slow descent using non-fossil fuels. (The latest model of the tether proposes "gates" such as the "Lunar Gate" or the "Mars Gate" as the optimal launch and return points for those destinations.)

Q: Why launch and return to specific gates; why not just launch to outer space destinations from the Apex Anchor for maximum velocity?

A: More fuel would be needed for slowing upon reaching your destination, but then, the Space Elevator would make the cost of getting fuel to space much more feasible. (Also, the farther out the elevator goes, the more we can charge for cargo and passengers!)

S: The tether from the Earth Node to the Apex Anchor and all points in-between amount to one large "Galactic Harbour."

Someone must have asked about locations for the Earth Node because that was the next topic. A presentation will be done by Vernon Hall on Sunday about this very topic, including input from the Master Thesis of Johan Fredrick Petersson suggesting Kiribati for the first port, then adding more around the globe, then eventually returning to the original location to convert it to a port for space tourism. Once in space, personnel (and cargo) could possibly transfer from tether to tether. This may generate competition and the possibility of using one tether to climb to a specific location, then transfer to a different tether because costs are lower on that tether in that region. This led to discussion of capitalism in space. Would a tourist grab a space "Uber" vehicle to get to his next destination? Would packages travel via Space Parcel Service? (When it absolutely HAS to get to Jupiter by next year...)

The next topic was regarding oversight. Which organization should be in charge of the Space Elevator? Should a new organization be formed, such as the "International Space Elevator Authority"? Whatever is decided, it will require international treaties or multinational agreements with U.N. oversight and perhaps FAA oversight. A U.N. based organization exists in Geneva named COPUOS (Committee on Peaceful Uses of Outer Space.)

Q: How are we going to get the funding for this? Will we need a bank at the Apex Anchor? It was brought up that the tether may produce electricity merely from the fields generated. This energy could be used to power the node, transfer power to vehicles, and even possibly be sold. There is the possibility that so much power will be generated that it will need to be discharged into space. It was suggested that a study be done and John Knapman brought up his study on Magneto Hydrodynamics.

Q: How would the tether be initially assembled/launched?

A: It is proposed that the components would be launched to LEO, assembled, then moved to GEO, where it would deploy towards space and earth to maintain equilibrium until it reaches both points.

- You don't have to grab on the tether to stay at the Geo. You just hang there without any energy consumption.
- Need to be make sure that the Geo Node stuff isn't too near the tether, to avoid collisions.
- Can look either up or down the tether, monitor the motions, perfect middle point
- Central command, shortest point between all points.
- Can have the greatest influence on tether dynamics at the bottom.
- Forces greatest at Geo, greatest load.
- Use taper of the tether to manage it.
- What kind of dampening? Lateral or up/down? Different approaches depending on which?
- Seems like Geo is a good place for lateral tether management.
- Relative station keeping? Formation flying? Relative position keeping. Swarm style. Maybe adapting to where the others are, not necessarily all going against a single reference point. Using rules for station keeping, not absolutes.
- How does the swam interface with the tether, climber, and/or payload?
- Need some sort of arm to reach out to climbers or payload, for off/on loading. Need to be able to quickly retract the arm in case it was necessary. "Pull off" style or "Put on" arm. Maybe not an arm, but a spring. Need to watch mass. Maybe magnets instead?
- Swam enabled you to be more efficient about what is dealt with and what isn't, and divide up responsibility among modules/nodes.
- Minimal node at Geo on the tether? Should it be surrounded by something?
- Ideal to have a system where no stopping of the climber is desired in most cases.
- Tether is widest at Geo, possible issues of slowing/speeding up, and exceeding stress/tensions locally in the area.
- Starting/stopping does affect lateral movement of the tether, may affect whether you can surround the tether or not.
- Possible change the material used in the tether past Geo, since gravity is no longer a factor. Maybe steel, or Kevlar.
- Geo node could beam power down or up for climbers or swarm nodes. Could have one central place at Geo that generates and stores power, and beams to others in the swarm. Or even beaming to other satellites in orbit.

- Solar power is nearly always present at Geo for usage.
- Study using the tether itself for power transmission.
- Solar sail as a means for swam nodes to move around. Possibly too slow for repositioning.
- Counterweights as a way to move up and down relatively small distances. As a way to reposition.
- Shock sensors on the climber to measure how much force is imparted to the payload as it went up.
- Need to keep track of everything that is happening along the entire Space Elevator system, save data/share data.
- Geo node could track a lost climber/payload.
- Be able to accept feedback and data from other sources in the "region" of Geo, other satellites.
- Geo watches other satellites' behavior, looking for problems/issues with them (orbit/plane changes), especially polar orbiting ones.
- Geo could observe space junk and track it also. Security and/or Surveillance functions.
- Geo is the right place for tether maintenance?

www.ingramcontent.com/pod-product-compliance
Lightning Source LLC
Chambersburg PA
CBHW080820170526
45158CB00009B/2478